Chemistry Research and Applications

Chemistry Research and Applications

What to Know about Lanthanum
Catherine C. Bradley (Editor)
2023. ISBN: 979-8-88697-615-1 (Softcover)
2023. ISBN: 979-8-88697-623-6 (eBook)

The Future of Biorefineries
Waldemar Nyström (Editor)
2023. ISBN: 979-8-88697-524-6 (Hardcover)
2023. ISBN: 979-8-88697-528-4 (eBook)

Properties and Uses of Antimony
David J. Jenkins (Editor)
2022. ISBN: 979-8-88697-081-4 (Softcover)
2022. ISBN: 979-8-88697-088-3 (eBook)

The Science of Carbamates
Güllü Kaymak (Editor)
2022. ISBN: 978-1-68507-708-2 (Softcover)
2022. ISBN: 978-1-68507-872-0 (eBook)

Deep Eutectic Solvents: Properties, Applications and Toxicity
Carlos Eduardo de Araújo Padilha, PhD, Everaldo Silvino dos Santos, PhD, Francisco Canindé de Sousa Júnior, PhD, Nathália Saraiva Rios, PhD (Editors)
2022. ISBN: 978-1-68507-719-8 (Hardcover)
2022. ISBN: 978-1-68507-799-0 (eBook)

Polycyclic Aromatic Hydrocarbons: Sources, Exposure and Health Effects
Warren L. Gregoire (Editor)
2022. ISBN: 978-1-68507-626-9 (Softcover)
2022. ISBN: 978-1-68507-685-6 (eBook)

More information about this series can be found at
https://novapublishers.com/product-category/series/chemistry-research-and-applications/

Roger G. Ward
Editor

Pyrimidines and Their Importance

Copyright © 2023 by Nova Science Publishers, Inc.

All rights reserved. No part of this book may be reproduced, stored in a retrieval system or transmitted in any form or by any means: electronic, electrostatic, magnetic, tape, mechanical photocopying, recording or otherwise without the written permission of the Publisher.

We have partnered with Copyright Clearance Center to make it easy for you to obtain permissions to reuse content from this publication. Simply navigate to this publication's page on Nova's website and locate the "Get Permission" button below the title description. This button is linked directly to the title's permission page on copyright.com. Alternatively, you can visit copyright.com and search by title, ISBN, or ISSN.

For further questions about using the service on copyright.com, please contact:
Copyright Clearance Center
Phone: +1-(978) 750-8400   Fax: +1-(978) 750-4470   E-mail: info@copyright.com

NOTICE TO THE READER

The Publisher has taken reasonable care in the preparation of this book, but makes no expressed or implied warranty of any kind and assumes no responsibility for any errors or omissions. No liability is assumed for incidental or consequential damages in connection with or arising out of information contained in this book. The Publisher shall not be liable for any special, consequential, or exemplary damages resulting, in whole or in part, from the readers' use of, or reliance upon, this material. Any parts of this book based on government reports are so indicated and copyright is claimed for those parts to the extent applicable to compilations of such works.

Independent verification should be sought for any data, advice or recommendations contained in this book. In addition, no responsibility is assumed by the Publisher for any injury and/or damage to persons or property arising from any methods, products, instructions, ideas or otherwise contained in this publication.

This publication is designed to provide accurate and authoritative information with regard to the subject matter covered herein. It is sold with the clear understanding that the Publisher is not engaged in rendering legal or any other professional services. If legal or any other expert assistance is required, the services of a competent person should be sought. FROM A DECLARATION OF PARTICIPANTS JOINTLY ADOPTED BY A COMMITTEE OF THE AMERICAN BAR ASSOCIATION AND A COMMITTEE OF PUBLISHERS.

Additional color graphics may be available in the e-book version of this book.

Library of Congress Cataloging-in-Publication Data

ISBN: 979-8-88697-656-4

Published by Nova Science Publishers, Inc. † New York

Contents

Preface vii

Chapter 1   **Synthesis, Physico-Chemical Properties and Antifungal Activity of New Hybrids of Thiazolo[4,5-D]Pyrimidines with (1h-1,2,4)Triazole** 1
S. Blokhina, A. Sharapova and M. Ol'khovich

Chapter 2   **Pyrimidine Ring Containing Natural Products and their Biological Importance** 57
Ravindra K. Rawal, Bishal Dutta and Pitambar Patel

Chapter 3   **Regulation of the Pyrimidine Biosynthetic Pathway in the Bacterium *Pseudomonas chlororaphis*** 109
Akram Bani Ahmad and Thomas P. West

Chapter 4   **Pyrimidines as Potential Corrosion Inhibitors: Recent Developments and Future Perspectives** 131
Dheeraj Singh Chauhan, M. A. Quraishi, Chandrabhan Verma and V. S. Saji

Index 155

Preface

This book contains four selected chapters on pyrimidines and their importance. The first chapter examines the synthesis, physico-chemical properties, and anti-fungal activity of new hybrids of thiazolo[4,5-D] pyrimidines with (1H-1,2, 4) triazole. The second chapter reviews pyrimidine ring containing natural products and their biological importance. The third chapter examines the regulation of the pyrimidine biosynthetic pathway in the bacterium Pseudomonas chlororaphis. The fourth and final chapter looks at pyrimidines as potential corrosion inhibitors.

Chapter 1 - A new series of hybrids of thiazolo[4,5-d]pyrimidines with (1H-1,2,4)triazole containing different linker skeletons and substituents (methyl-, methoxy-, chloro- and fluoro-) in the phenyl ring were designed and synthesized as antifungal agents. The chemical structures of the target compounds were characterized by physico-chemical (1H-NMR, LC-MS) methods. The microdilution broth method was used to investigate their antifungal activities. Structure-activity relationship studies indicated that the introduction of piperazine into the linker is crucial to antifungal activity. The final compounds exhibited in vitro antifungal activity against the following filamentous and yeast fungi: C. parapsilosis ATCC 22019, C. albicans ATCC 24433, C. albicans 8R, C. albicans CBS 8837, C. albicans 604M, C. utilis 84, C. tropicalis 3019, C. glabrata 61 L, C. krusei 432M, Cryptococcus neoformans, A. niger 37a, M. canis B-200, T. rubrum 2002 comparable to fluconazole. The amorphous nature of the obtained compounds was verified by DSC and PXRD. The pharmaceutically significant properties of the obtained substances – solubility and lipophilicity – were determined. The kinetic and equilibrium solubilities of the derivatives were studied by the shake-flask method in pharmaceutically relevant solvents: buffer pH 2.0, buffer pH 7.4 and 1-octanol. The solubility of the compounds within the temperature range of (293.15 −313.15) K was found not to exceed $8.6·10^{-4}$ mol·L-1 in aqueous solvents and $6·10^{-2}$ mol·L-1 in alcohol. The solubility values were correlated by the van't Hoff and modified Apelblat equations.

Based on the partition coefficients in the 1-octanol/buffer pH 7.4 system the lipophilicity of the compounds was evaluated. The thermodynamic functions of dissolution and partition for the studied compounds in the selected solvents were calculated. The process of transfer of the compounds from the aqueous phase to the organic one was found to be endothermic and entropy-controlled. The efficiency of some computer programs to predict partition coefficients was tested using experimental data. The in vitro antifungal activity of the studied compounds was shown to increase as their solubility in aqueous solvents became higher.

Chapter 2 - Natural products remain an important source as well as inspiration for modern drug discovery. Among the natural products, heterocyclic compounds are at the forefront of drug discovery because of their high structural diversity, vast biological activities, and toxicity. In the family of heterocyclic natural products, the alkaloid class of natural products is the crucial group of compounds that are widely used in medicinal chemistry and drug discovery. However, this book chapter is mainly focused on the pyrimidine ring containing natural products and their biological activities along with clinical and *in vitro* applications. Substituted pyrimidine ring is widely found in many bioactive natural products such as vitamins, health supplements, antiviral, antifungal, antibacterial, etc. Based on their structural features, the pyrimidine class of natural products is divided into five main groups namely; i) tethered pyrimidines, ii) fused pyrimidines, iii) pyrimidine nucleoside and nucleotides and iv) vitamins and v) toxins. Each group is further divided into sub-groups based on the nature of substituents and their relative biological activities are discussed in detail.

Chapter 3 - The pyrimidine biosynthetic pathway consists of five enzymes that are unique to the formation of pyrimidine nucleotides. The pyrimidine biosynthetic pathway enzymes are aspartate transcarbamoylase, dihydroorotase, dihydroorotate dehydrogenase, orotate phosphoribosyltransferase and orotidine 5'-monophosphate decarboxylase. The focus of this study was to explore the regulation of pyrimidine biosynthesis in *Pseudomonas chlororaphis* ATCC 17414. To do so, the effect of supplementing the pyrimidine bases orotic acid or uracil into a glucose or succinate-containing culture medium of *P. chlororaphis* ATCC 17414 on the pyrimidine biosynthetic enzyme activities was first explored. Transcarbamoylase and decarboxylase activities were repressed by orotic acid or uracil addition independent of carbon source in the ATCC 17414 cells. Next, a pyrimidine auxotrophic mutant of *P. chlororaphis* was isolated using chemical mutagenesis and 5-fluororotic acid resistance. The isolated mutant utilized either uracil, uridine or cytosine as a pyrimidine source to support their growth. The mutant strain was deficient for orotidine 5'-monophosphate decarboxylase activity. The mutant cells were subjected to pyrimidine limitation to determine if nucleotide depletion influenced the synthesis of the pyrimidine biosynthetic enzyme activities. The pyrimidine limitation of a pyrimidine auxotrophic strain often causes derepression of the synthesis of the pyrimidine pathway enzymes. When the orotidine 5'-monophosphate decarboxylase mutant strain cells were limited for pyrimidines for one hour, transcarbamoylase, dihydroorotase, dehydrogenase and phosphoribosyltransferase activities were all derepressed compared to their activities in the mutant strain grown in excess uracil. To learn if the initial pathway enzyme aspartate transcarbamoylase was subject to regulation of its activity, the effect of pyrophosphate and ribonucleotides in *P. chlororaphis* ATCC 17414 cells was studied. The transcarbamoylase was highly inhibited by uridine 5'-monophosphate, uridine 5'-diphosphate, cytidine 5'-monophosphate and guanosine 5'-triphosphate in the glucose-grown ATCC 17414 cells. Overall, it was concluded that pyrimidine biosynthetic pathway enzymes in *P. chlororaphis* ATCC 17414 was regulated at the level of enzyme synthesis and that the initial pathway enzyme aspartate transcarbamoylase activity was controlled at the level of enzyme activity.

Chapter 4 - Heterocyclic molecules containing nitrogen are extensively studied as corrosion inhibitors due to their capability to form strong coordination bonds with metallic atoms. The excellent corrosion protection action of pyrimidine derivatives can be attributed to the presence of electron-withdrawing and electron-donating substituents that allow an improved protection performance compared to that of the parent pyrimidine molecule. Computational studies have provided evidence that the pyrimidine-based molecules undergo donor-acceptor type interactions with metallic substrates. In general, the organic corrosion inhibitors undergo a mixed-type of adsorption behavior, and their adsorption follows the Langmuir isotherm. This chapter presents an overview of pyrimidine-based molecules as corrosion inhibitors for metals and alloys in a variety of corrosive environments.

Chapter 1

Synthesis, Physico-Chemical Properties and Antifungal Activity of New Hybrids of Thiazolo[4,5-D]Pyrimidines with (1h-1,2,4)Triazole

S. Blokhina, PhD
A. Sharapova[*], PhD
and M. Ol'khovich, PhD

G. A. Krestov Institute of Solution Chemistry, Russian Academy of Sciences, Ivanovo, Russian Federation

Abstract

A new series of hybrids of thiazolo[4,5-d]pyrimidines with (1H-1,2,4)triazole containing different linker skeletons and substituents (methyl-, methoxy-, chloro- and fluoro-) in the phenyl ring were designed and synthesized as antifungal agents. The chemical structures of the target compounds were characterized by physico-chemical (^{1}H-NMR, LC-MS) methods. The microdilution broth method was used to investigate their antifungal activities. Structure-activity relationship studies indicated that the introduction of piperazine into the linker is crucial to antifungal activity. The final compounds exhibited in vitro antifungal activity against the following filamentous and yeast fungi: *C. parapsilosis* ATCC 22019, *C. albicans* ATCC 24433, *C. albicans* 8R, *C. albicans* CBS 8837, *C. albicans* 604M, *C. utilis* 84, *C. tropicalis* 3019, *C. glabrata* 61 L, *C. krusei* 432M, *Cryptococcus neoformans*, *A. niger* 37a, *M. canis* B-200, *T. rubrum* 2002 comparable to fluconazole. The

[*] Corresponding Author's Email: avs@isc-ras.ru.

In: Pyrimidines and Their Importance
Editor: Roger G. Ward
ISBN: 979-8-88697-656-4
© 2023 Nova Science Publishers, Inc.

amorphous nature of the obtained compounds was verified by DSC and PXRD. The pharmaceutically significant properties of the obtained substances – solubility and lipophilicity – were determined. The kinetic and equilibrium solubilities of the derivatives were studied by the shake-flask method in pharmaceutically relevant solvents: buffer pH 2.0, buffer pH 7.4 and 1-octanol. The solubility of the compounds within the temperature range of (293.15 –313.15) K was found not to exceed $8.6 \cdot 10^{-4}$ mol·L^{-1} in aqueous solvents and $6 \cdot 10^{-2}$ mol·L^{-1} in alcohol. The solubility values were correlated by the van't Hoff and modified Apelblat equations. Based on the partition coefficients in the 1-octanol/buffer pH 7.4 system the lipophilicity of the compounds was evaluated. The thermodynamic functions of dissolution and partition for the studied compounds in the selected solvents were calculated. The process of transfer of the compounds from the aqueous phase to the organic one was found to be endothermic and entropy-controlled. The efficiency of some computer programs to predict partition coefficients was tested using experimental data. The in vitro antifungal activity of the studied compounds was shown to increase as their solubility in aqueous solvents became higher.

Keywords: hybrids, synthesis, solubility, lipophilicity, antifungal activity

Introduction

The last few decades have seen a considerable increase in fungal diseases and emergence of resistance to the most popular antifungal drugs belonging to a lot of chemical classes, which makes treatment of patients more difficult. Fungal infections are especially dangerous in patients with a weakened immune system destroyed by cancer, HIV, immunosuppressive drugs or other diseases (Perlin, Shor, and Zhao, 2015; Hsu et al., 2011). The enormous amount of available data on the chemical characteristics and biological activity of (1H-1,2,4)triazole derivatives allows researchers to consider these substances to be one of the most promising classes of drug compounds (Zhou and Wang, 2012; Emami et al., 2019).

Some of the (1H-1,2,4)triazole derivatives exhibit antibacterial, analeptic, anti-inflammatory, cardioprotective and other types of activity (Zirngibl, 1998). Common antifungal drugs - fluconazole, itraconazole, isavuconazole, efinaconazole, etc. - include a triazole fragment in the structure of their molecules. The mechanism of azole action is associated with inhibiting P450 cytochrome enzymes catalyzing lanosterol transformation into ergosterol, which leads to the destruction of the fungus cell membrane (Egbuta, Lo, and Ghosh, 2014).

Synthesis of antifungal drugs by the method of binding pharmacophore heterocycles with a linker is one of the largest branches of pharmaceutical chemistry. A lot of works have described attempts to find the best combination with a triazole fragment by changing the molecule components (Bozorov, Zhao, and Aisa, 2019; Aggarwal and Sumran, 2020). This principle has been used to obtain a lot of drugs of the triazole group, including ravuconazole and its commercial water-soluble analogue - isavuconazole - representing a triazole hybrid with thiazole (Slavin and Thursky, 2016). Sulfur-containing heterocycles are often used by researchers as one of the components of the hybrid molecule based on triazole. An example would be the tetrahydro-thiazolo-pyridine hybrids with high antifungal activity and water solubility (Chai et al., 2009; Sheng et al., 2010). The effectiveness of the action of the hybrids with various thienopyrimidones was demonstrated on experimental models of systemic candida infections (Bartolli et al., 1998). The high activity indicator was also achieved in the tests of various condensed pyrimidones with benzothiazoles (Guillon et al., 2013).

One of the main trends in medicinal chemistry is designing broad-spectrum drug compounds by combining various pharmacophore groups and fragments in one molecule (Bérubé et al., 2016). One of such studies is described in the present work, where thiazolo[4,5-d]pyrimidine was used as a component of the hybrid molecule. The thiazolo[4,5-d]pyrimidine derivatives potentially have biological relevance as analgesics, central nervous system depressants, and bronchodilators and are also recommended for prevention and treatment of atherosclerosis, diabetes and other diseases (El-Bayouki and Basyouni, 2010; Kuppast and Fahmy, 2016). Another important fragment included in a lot of drugs is piperazine (Brito et al., 2019; Rathi et al., 2016). There are a lot of examples in the literature, in which piperazine is combined with a variety of heterocyclic fragments significantly improving the biological properties of the compounds, including their antimicrobial activity (Emami et al., 2019; Heeres, Meerpoel, and Levi, 2010).

Molecular structure is decisive for the application of bioactive compounds in drugs since it is responsible for their ADME properties (absorption, distribution, metabolism, and excretion). This set of fundamental physico-chemical and biochemical properties of drugs largely depends on their solubility in physiological media and lipophilicity (Daina, Michielin, and Zoete, 2017; Brittain, 1995). The solubility of bioactive compounds is determined with the help of model solvents represented by aqueous solutions:

buffer pH 2.0 imitating the gastro-intestinal tract environment and buffer pH 7.4 modeling the blood plasma. The most important organic solvent is 1-octanol whose amphiphilic properties allow its application as a model of cell phospholipid membranes. Consequently, the partition coefficient in the 1-octanol / water system is an indicator of the lipophilicity of a compound (Sangster, 1997). Among the techniques currently used for screening membrane permeability are the Caco-2 monolayer model, the PAMPA (parallel artificial membrane permeation assay) and the ILG (immobilized liposome chromatography) methods (Kansy, Avdeef and Fischer, 2004). These methodologies make it possible to more adequately simulate the drug transport process in comparison with the 1-octanol / water system model. However, the 1-octanol/water system is widely used to study lipophilicity of bioactive compounds, providing simplicity of the experiment and giving well reproducible results (Kempinska et al., 2019). In addition, based on the previously established correlation of drug partition in the 1-octanol/water system with biological activity, this parameter is used as a key descriptor in the search for the relationship between the quantitative structure and activity (QSAR) (Noorizadeh, Sajjadifar, and Sobhanardakani, 2014).

Now one of the main drawbacks of the potential new drug compounds is their low solubility in water (Buckley et al., 2013). In this regard, there is an increasing interest in developing approaches to improve the solubility of these drugs: microcapsule, liposome, suspension with nanosphere, cyclodextrin inclusion complex, cocrystal and solid dispersion techniques (Shah, Goodyear, and Michniak-Kohn, 2017). Using amorphous drug forms instead of their crystalline analogues is also an effective method of improving their solubility and dissolution rate (Kalepu and Nekkanti, 2015; Bergström, and Avdeef, 2019). And it should be taken into account that a compound in its crystalline and amorphous forms has different physical, chemical and mechanical properties, which affects the drug stability and bioavailability (Blagden, et al., 2007; Franks, 1993).

In this work, we obtained compounds consisting of two components specified earlier: (1H-1,2,4)triazole, thiazolo-pyrimidine combining them into one hybrid molecular structure. We also studied the effect of the structure of the linker between the heterocycles and substituents of various chemical nature on the in vitro antifungal activity of the synthesized substances.

We studied the solubility in pharmaceutically relevant solvents and lipophilicity and bioactivity of the novel potential antifungal thiazolo[4,5-d]pyrimidine derivatives. The studies of the intermolecular interactions of the drug compounds in pharmaceutically relevant aqueous and organic solvents make it possible to optimize the selection procedure of promising compounds at the in vitro stage before conducting expensive in vivo tests.

Experimental Section

General

Bidistilled water (with the electrical conductivity of 2.1 $\mu S\ cm^{-1}$) was used to prepare the buffer solutions. Phosphate buffer pH 7.4 (I=0.15 $mol \cdot L^{-1}$) was prepared by combining the $KHPO_4$ (9.1 g in 1 L) and $NaH_2PO_4 \cdot 12H_2O$ (23.6 g in 1 L) salts. The pH values were measured by using an FG2-Kit pH meter (Mettler Toledo, Switzerland) standardized with pH 1.68, 6.86 and 9.22 solutions. All the other chemicals and reagents were of analytical grade.

Synthesis

General Procedure for Compounds of Group 1

Equimolar amounts of isothiocyanates were added to a mixture of 2-cyanoacetamide, finely ground sulfur and triethylamine in ethanol. The reaction mixture was boiled for 1 h. After it was cooled to room temperature, the precipitate was filtered and washed on a filter with cold ethanol. The respective substituted 2-thioxo-dihydro-thiazolo-carboxamide obtained by this method was mixed with an excess of triethyl orthoformate (3 equiv.) and dissolved in acetic oxide. After the reaction mixture was stored at room temperature for half an hour, it was boiled for several hours. The precipitate of the respective 2,3-dihydrothiazolo[4,5-d]pyrimidin-7-ones obtained after cooling was filtered and used without additional purification.

Synthesis of 6-[2-(2,4-difluorophenyl)-2-hydroxy-3-(1H-1,2,4-triazol-1-yl)propyl]-3-(p-tolyl)-2-thioxo-thiazolo[4,5-d]pyrimidin-7-one (**1a**).

0.086 g (0.36 mmol) 1-((2-(2,4-difluorophenyl)oxirane-2-yl)methyl)-(1H-1,2,4)triazole, 0.1 g (0.36 mmol) 2-thioxo-3-(p-tolyl)-2,3-dihydrothiazolo[4,5-d]pyrimidine-7(6H)-one, 0.1 g (0.72 mmol) potassium carbonate and 0.019 g (0.36 mmol) ammonium chloride (at a temperature of 120 oC) were heated and mixed in 3 mL of N,N-dimethyl formamide for 36 hours. After cooling the reaction mixture to room temperature, we added 10 mL water and subjected the mixture to extraction with ethylacetate (3x30 mL).

The combined organic phase was washed with a saturated NaCl solution (50 mL), dried over sodium sulfate and evaporated. The target compound was purified by column chromatography (with a petroleum ether: ethylacetate eluent). The yield was 0.077 g (41%), yellow powder. $C_{23}H_{18}F_2N_6O_2S_2$. m/z 512,565. LCMS [M+1]$^+$ = 513,1 $^+$

^{1}H NMR (DMSO-d_6, δ, ppm, J/Hz): 2.41 (s, 3 H, CH$_3$), 4.32 (d, 1 H, $^3J_{HH}$= 14.0, CH$_2$), 4.50 (d, 1 H, $^3J_{HH}$= 14.4, CH$_2$), 4.65 (d, 1 H, $^3J_{HH}$= 14.0, CH$_2$), 4.81 (d, 1 H, $^3J_{HH}$= 14.5, CH$_2$), 6.37 (s, 1 H, aromatic), 6.92 (td, 1 H, $^3J_{HH}$= 8.5, 1.6, aromatic), 7.17 - 7.24 (m, 1 H, aromatic), 7.28 (d, 2 H, $^3J_{HH}$= 8.1, aromatic), 7.40 (d, 2 H, $^3J_{HH}$= 8.2, aromatic), 7.71 (s, 1 H, N=CH-N), 8.24 (s, 1 H, N=CH-N), 8.29 (s, 1 H, N=CH-N).

Compounds **1b, c** were obtained similarly by the general method.

Synthesis of 6-[2-(2,4-difluorophenyl)-2-hydroxy-3-(1H-1,2,4-triazol-1-yl)propyl]-3-(2-methoxyphenyl)-2-thioxo-thiazolo[4,5-d]pyrimidin-7-one (**1b**). The yield was 0.101 g (37%), yellow powder. $C_{23}H_{18}F_2N_6O_3S_2$. m/z 528,564. LCMS [M+1]$^+$ = 529.0 $^+$.

^{1}H NMR (DMSO-d_6, δ, ppm, J/Hz): 3.74 (s, 3 H, OCH$_3$), 4.39 (d, 1 H, $^3J_{HH}$= 14.1, CH$_2$), 4.49 (d, 1 H, $^3J_{HH}$= 14.4, CH$_2$), 4.67 (d, 1 H, $^3J_{HH}$= 13.9, CH$_2$), 4.82 (d, 1 H, $^3J_{HH}$= 14.7, CH$_2$), 6.40 (c, 1 H, aromatic), 6.91 – 6.94 (m, 1 H, aromatic), 7.12 - 7.17 (m, 2 H, aromatic), 7.27 – 7.29 (m, 1 H, aromatic), 7.33 - 7.35 (m, 1 H, aromatic), 7.53 – 7.56 (m, 1 H, aromatic), 7.72 (s, 1 H, N=CH-N), 8.26 (s, 1 H, N=CH-N), 8.28 (s, 1 H, N=CH-N).

Synthesis of 6-[2-(2,4-difluorophenyl)-2-hydroxy-3-(1H-1,2,4-triazol-1-yl)propyl]-3-(4-fluorophenyl)-2-thioxo-thiazolo[4,5-d]pyrimidin-7-one (**1c**). The yield was 0.088 g (24%), yellow powder. $C_{22}H_{15}F_3N_6O_2S_2$. m/z 516,526. LCMS [M+1] $^+$ = 517 $^{+1}$.

^{1}H NMR spectrum (DMSO-d_6, δ, ppm, J/Hz): 4.35 (d, 1 H, $^3J_{HH}$= 13.9, CH$_2$), 4.49 (d, 1 H, $^3J_{HH}$= 14.5, CH$_2$), 4.64 (d, 1 H, $^3J_{HH}$= 14.0, CH$_2$), 4.82 (d, 1 H, $^3J_{HH}$= 14.6, CH$_2$), 6.39 (c, 1 H, aromatic), 6.90 – 6.93 (m, 1 H, aromatic), 7.23 - 7.27 (m, 1 H, aromatic), 7.44 (d, 2 H, $^3J_{HH}$= 8.8, aromatic), 7.51 (d, 2 H, $^3J_{HH}$= 8.74, aromatic), 7.72 (s, 1 H, N=CH-N), 8.25 (s, 1 H, N=CH-N), 8.31 (s, 1 H, N=CH-N).

General Procedure for Compounds of Group 2

We dissolved equivalent amounts of 2-thioxo-3-(R1,R2=phenyl)-2,3-dihydrothiazolo[4,5-d]pyrimidin-7(6H)-one and 1-(2-chloroacetyl amino)-2-(2,4-difluorophenyl)-3-(1H-1,2,4-triazol-1-yl)propan-2-ol in DMSO and added K$_2$CO$_3$ (1 equiv.) and KI (1 equiv.) to the obtained mixture. The reaction mixture was stirred for 7 hours and simultaneously heated to 85 °C. After that, 10 mL of water were added and then the mixture was extracted by ethylacetate. The obtained organic phase was washed with a saturated NaCl solution (50 mL), dried over sodium sulfate and evaporated. The target 6-[2-(2,4-difluorophenyl)-2-hydroxy-3-(1H-1,2,4-triazol-1-yl)propyl]-2-[3-(R1,R2-phenyl)-7-oxo-2-thioxo-thiazolo[4,5-d]pyrimidin-6-yl]acetamides (**2a - d**) were obtained and purified by column chromatography (with a petroleum ether:ethyl acetate eluent).

Synthesis of 6-[2-(2,4-difluorophenyl)-2-hydroxy-3-(1H-1,2,4-triazol-1-yl)propyl]-2-[7-oxo-3-(p-tolyl)-2-thioxo-thiazolo[4,5-d]pyrimidin-6-yl]acetamide (**2a**). (R_1 = CH_3, R_2 = H). Yield: 69%, yellow powder. $C_{25}H_{21}F_2N_7O_3S_2$. m/e 569.60. LCMS [M+1]$^+$ = 570 $^{+1}$.

^{1}H NMR spectrum (DMSO-d$_6$, δ, ppm, J/Hz): 2.40 (s, 3 H, CH_3), 3.53 (dd, 1 H, $^3J_{HH}$ = 14.0, 5.6, CH_2), 3.74 (dd, 1 H, $^3J_{HH}$ = 13.7, 6.3, CH_2), 4.50 (d, 1 H, $^3J_{HH}$ = 14.4, CH_2), 4.57 - 4.71 (m, 3 H, CH_2), 6.95 (td, $^3J_{HH}$ = 8.4, 2.5, 1 H, aromatic), 7.11 – 7.20 (m, 1 H, aromatic), 7.30 (d, 2 H, $^3J_{HH}$ = 8.3, aromatic), 7.33 - 7.43 (m, 3 H, aromatic), 7.75 (s, 1 H, N=CH-N), 8.27 (s, 1 H, N=CH-N), 8.32 (s, 1 H, N=CH-N).

Synthesis of 6-[2-(2,4-difluorophenyl)-2-hydroxy-3-(1H-1,2,4-triazol-1-yl)propyl]-2-[3-(2-methoxyphenyl)-7-oxo-2-thioxo-thiazolo[4,5-d]pyrimidin-6-yl]acetamide (**2b**). (R_1 = H, R_2 = OCH_3). Yield: 55%, yellow powder. $C_{25}H_{21}F_2N_7O_4S_2$. m/e 585.61. LCMS [M+1]$^+$ = 586.0 $^{+1}$.

^{1}H NMR spectrum (DMSO-d$_6$, δ, ppm, J/Hz): 3.51 (dt, 1 H, $^3J_{HH}$ = 13.7, 4.7, CH_2), 3.71 - 3.79 (m, 4 H, CH_2, OCH_3), 4.50 (d, 1 H, $^3J_{HH}$ = 14.4, CH_2), 4.58 - 4.73 (m, 3 H, CH_2), 6.95 (td, 1 H, $^3J_{HH}$ = 8.5, 2.1, aromatic), 7.11 - 7.20 (m, 2 H, aromatic), 7.28 (d, 1 H, $^3J_{HH}$ = 8.4, aromatic), 7.31 - 7.40 (m, 2 H, aromatic), 7.52 - 7.59 (m, 1 H, aromatic), 7.74 (s, 1 H, N=CH-N), 8.27 (s, 1 H, N=CH-N), 8.31 (s, 1 H, N=CH-N).

Synthesis of 6-[2-(2,4-difluorophenyl)-2-hydroxy-3-(1H-1,2,4-triazol-1-yl)propyl]-2-[3-(4-fluorophenyl)-7-oxo-2-thioxo-thiazolo[4,5-d]pyrimidin-6-yl]acetamide (**2c**). (R_1 = F, R_2 = H). Yield: 80%, yellow powder. $C_{24}H_{18}F_3N_7O_3S_2$. m/e 573.57. LCMS [M+1]$^+$ = 574 $^{+1}$.

^{1}H NMR spectrum (DMSO-d$_6$, δ, ppm, J/Hz): 3.53 (dd, 1 H, $^3J_{HH}$ = 13.9, 5.6, CH_2), 3.74 (dd, 1 H, $^3J_{HH}$ = 13.8, 6.2, CH_2), 4.50 (d, 1 H, $^3J_{HH}$ = 14.4, CH_2), 4.58 - 4.73 (m, 3 H, CH_2), 6.95 (td, 1 H, 3JHH = 8.5, 2.4, aromatic), 7.16 (ddd, 1 H, $^3J_{HH}$ = 11.9, 9.3, 2.5, aromatic), 7.36 (td, 1 H, $^3J_{HH}$ = 8.9, 7.0, aromatic), 7.42 – 7.47 (m, 2 H, aromatic), 7.51 - 7.56 (m, 2 H, aromatic), 7.74 (s, 1 H, N=CH-N), 8.27 (s, 1 H, N=CH-N), 8.34 (s, 1 H, N=CH-N).

Synthesis of 6-[2-(2,4-difluorophenyl)-2-hydroxy-3-(1H-1,2,4-triazol-1-yl)propyl]-2-[3-(4-chlorophenyl)-7-oxo-2-thioxo-thiazolo[4,5-d]pyrimidin-6-yl]- acetamide (**2d**). (R_1 = Cl, R_2 = H). Yield: 73%, yellow powder. $C_{24}H_{18}ClF_2N_7O_3S_2$. m/e 590.02. LCMS [M+1]$^+$ = 591 [+1].

^{1}H NMR spectrum (DMSO-d_6, δ, ppm, *J*/Hz): 3.53 (dd, 1 H, $^3J_{HH}$ = 13.8, 5.6, CH_2), 3.74 (dd, 1 H, $^3J_{HH}$ = 13.8, 6.3, CH_2), 4.00 (d, 1 H, $^3J_{HH}$ = 14.4, CH_2), 4.58 - 4.72 (m, 3 H), 6.95 (td, 1 H, $^3J_{HH}$ = 8.5, 2.5, aromatic), 7.16 (ddd, 1 H, $^3J_{HH}$ = 12.0, 9.3, 2.5, aromatic), 7.31 - 7.40 (m, 1 H, aromatic), 7.52 (d, 2 H, $^3J_{HH}$ = 8.7, aromatic), 7.68 (d, 2 H, $^3J_{HH}$ = 8.7, aromatic), 7.74 (s, 1 H, N=CH-N), 8.27 (s, 1 H, N=CH-N), 8.35 (s, 1 H, N=CH-N).

Biological Study

The activity evaluation was carried out in accordance with the recommendations of the Clinical and Laboratory Standards Institute (CLSI), a world non-for-profit medical standards development organization.

To determine the minimal inhibitory concentration (MIC), we used the micromethod of serial dilution in an RPMI 1640 medium (with glutamine produced by PanEco) with a 2% glucose addition. The following reference strains were analysed: *C. parapsilosis* ATCC 22019, *C. albicans* ATCC 24433, *C. albicans* 8R, *C. albicans* CBS 8837, *C. albicans* 604M, *C. utilis* 84, *C. tropicalis* 3019, *C. glabrata* 61L, *Cryptococcus neoformans*, and *C. krusei* 432M *yeast cultures, A.niger* 37a, *filamentous fungi M. canis* B-200 *and T.rubrum* 2002.

To prepare the base solutions with a concentration of 10.000 mg/L, the samples were dissolved in dimethyl sulfoxide. To prepare the working solutions with a concentration of 64 µg/ml, we brought the base solutions of 0.064 ml up to the volume of 10 ml by mixing them with an RPMI 1640 nutrient broth with a 2% glucose addition, with the final DMSO concentration not exceeding 0.3%.

The test-microorganisms were kept in low-temperature conditions (-75°C) in trypticase-soy broth with a 10-15% glycerol addition. To prepare the seed stock, the strains were grown in a Sabouraud agar medium (GRM 2 VFS 42-3068-98, Biohold, Russia) at 35°C: *Candida spp.* for 48 hours, filamentous fungi for about 2 weeks.

The *Candida spp.* seed suspension was prepared in a phosphate-buffered saline by the McFarland turbidity standard (~5 10^6 CFU/mL for the yeast cultures), estimated densimetrically (Den-1, Biosan), diluted 1:1000 until ~5

10^3 CFU/mL in an RPMI medium with a 2% glucose addition. For each test-strain of the filamentous fungi, the inoculum was prepared by abrading part of the colony in a physiological solution in a glass bead vial. The conidia and spores were collected with a pipette through a gauze filter, the calculation was made with a Goryaev chamber. Every inoculum was brought up to the working titer volume in an RPMI medium. The resulting suspension contained 1.5 - 3.2 10^4 CFU/mL. To control the titer of the viable colony-forming units (CFU), we transferred 10 mcL of the inoculum onto the Sabouraud agar medium.

In this work, we used 96-well plates for the immunological studies (Medpolymer, Saint Petersburg). We placed 100 mcL of a yeast culture suspension in the nutrient medium and the samples into the plate wells. The component concentrations ranged from 32 to 0.25 µg/ml. Each of the samples was analysed three times.

To make accurate measurements of the minimum inhibitory activity (MIC), we used standard Fluconazole samples (Sigma-Aldrich) as the internal standard. For growth control; all the test-cultures were inoculated into a nutrient medium without the samples.

The sensitivity was evaluated visually, after the incubation at 35 °C for 24 and 48 h for *Candida spp.* and 48-96 h for *A. niger, M. canis B, T. rubrum*, and the value was compared with the growth density in the reference culture without the samples.

Differential Scanning Calorimetry (DSC)

The thermophysical properties of the synthesized compounds were investigated using a Perkin-Elmer Pyris 1 DSC differential scanning calorimeter (Perkin-Elmer Analytical Instruments, Norwalk, Connecticut, USA) with Pyris software for Windows NT. The DSC runs were performed in an atmosphere of flowing 20 $cm^3 \cdot min^{-1}$ dry helium gas of high purity 0.99996 (mass fraction) using standard aluminum sample pans and a heating rate of 10 $K \cdot min^{-1}$. The weight measurement accuracy was 0.005 mg. The DSC was calibrated with an indium sample from Perkin-Elmer (P/N 0319-0033). The value determined for the enthalpy of fusion corresponded to 28.48 $J \cdot g^{-1}$ (the reference value being 28.45 $J \cdot g^{-1}$). The fusion temperature was 429.5 ± 0.1 K (determined from at least ten measurements).

Powder X-Ray Diffraction (PXRD)

The powder X-ray diffraction patterns of the compounds were recorded under ambient conditions in the Bragg-Brentano geometry with a Bruker D8 Advance diffractometer with CuK$_\alpha$ radiation (λ=1.5406 Å) at 40 kV and 40 mA power. The X-ray diffraction patterns were collected over the 2θ range of 5–30°, with a 0.03° step size.

Kinetic Solubility Experiment

The dissolution experiments were made by the shake-flask method in buffers pH 2.0 and 7.4 at 298.15 K for 4 days. An excess amount of the sample was suspended in the respective buffer solution in Pyrexglass tubes. The amount of the dissolved sample was measured by taking aliquots of the media at pre-determined time points. The suspension was filtered through a MILLEX®HA, 0.45 μm (Ireland), and the concentration was determined using a Cary-50 spectrophotometer (Varian, USA). Each experiment was repeated in triplicate. The equilibrium solid phases were characterized by PXRD.

Equilibrium Solubility

All the experiments were carried out by the isothermal saturation method at five temperature points: 293.15, 298.15, 303.15, 308.15, and 313.15 ± 0.1 K. The essence of the above method includes determination of the compound concentration in a saturated solution. Glass ampoules containing the tested substance and the solvent were placed into an air thermostat equipped with a stirring device. The time required to reach a constant value of the solution concentration was determined from the solubility kinetic dependences and averaged 4-5 days. The time of solid phase sedimentation after stirring was 2 hours. The solution aliquot was taken and centrifuged in a Biofuge stratos centrifuge (Germany) with temperature control for 5 minutes at a fixed temperature. The solid phase was removed by isothermal filtration with a 0.45 μm MILLEX®HA filter (Ireland). The saturated solution was diluted with the correspondent solvent to the required concentration. The molar solubilities of the drugs were measured by a Cary-50 spectrophotometer (Varian, USA) with an accuracy of 2 – 4%. The experimental results are reported as an average value of at least three replicated experiments. It should be noted that the sediment DSC analysis showed the absence of crystallosolvates in all the tested compounds.

The wavelengths corresponding to the absorption maximums (maxima?) for the compounds in the studied solvents were specified as 335 nm in buffer solutions (pH 7.4 and 2.0) and as 337 nm in 1-octanol. The calibration was made at room temperature using solutions with the known substance concentrations in each of the investigated solvents. The solutions were prepared by adding an appropriate mass of the substance and volume of the solvent (buffer pH 2.0, buffer pH 7.4 and 1-octanol) to the flask and mixing them until the substance was totally dissolved.

The conversion of molarity to the mole fraction concentration scale was made using equation (1):

$$x = \frac{M_2 S}{S(M_2 - M_1) + 1000\rho}, \qquad (1)$$

where S is the molarity (mol·L^{-1}), M_1 and M_2 are the molar masses of the solute and solvent, respectively, and ρ (g·cm^{-3}) is the density of the pure solvents. It should be noted that the solubility of the compounds studied is low; therefore the density of the saturated solutions is almost the same as the one of the pure solvent. The mole fraction solubilities for the buffer solutions were calculated taking into account the buffer compositions.

Thermodynamic Models

In this work, the van't Hoff and the Apelblat models are manipulated to describe the experimental solubility data.

Van't Hoff Equation

The Van't hoff equation is a universal equation which correlates the solute solubility in a real solution with variations in temperature. It explains the effect of temperature on the solubility of triazole derivatives. The solubility of the compounds under study calculated using the Van't Hoff model is as per Eq. (2).

$$\ln x = A + \frac{B}{(T/K)} \quad (2)$$

where x represents the mole fraction solubility of the triazole derivatives, T is the absolute temperature and A and B are the regression parameters obtained by the multivariable least-square method.

Modified Apelblat Equation

This empirical model shows the effect of temperature on the experimental mole fraction solubility of a solute in mono and binary solvents. The equation is derived from the Clausius–Clapeyron equation and shows the temperature effect on the enthalpy of solution. The modified Apelblat equation is given in Eq. (3).

$$\ln x = A + \frac{B}{T/K} + C\ln(T/K) \quad (3)$$

where x is the experimental saturated mole fraction solubility of the solute, T is the absolute temperature, A, B and C are the empirical model parameters. Values A and B represent the variation in the solution behavior resulting from the non-ideality of the solute solubility, whereas the value of C represents the association between the temperatures and the enthalpy of fusion.

Model Accuracy

The model accuracy can be predicted using *RD*, *RAD*, *RMSD* to calculate the difference between the experimental and predicted values of the compounds solubility. They can be obtained by Eq. (4 - 6).

The relative deviations between the experimental and calculated values (RD) are found according to

$$RD = (x_{exp} - x_{cal})/x_{exp} \quad (4)$$

The relative average deviations (RAD) are calculated according to:

$$RAD = \frac{1}{N}\sum_{i=1}^{N}\left|\frac{x_{exp} - x_{cal}}{x_{exp}}\right| \qquad (5)$$

The root-mean-square deviation (RMSD) is expressed as

$$RMSD = \left|\frac{1}{N}\sum_{i=1}^{N}(x_{exp} - x_{cal})^2\right|^{1/2} \qquad (6)$$

where N represents the number of experimental points, x_{exp} and x_{cal} are the experimental and calculated mole fraction solubility values of the compound, respectively.

Thermodynamic Framework of Dissolution

Thermodynamic functions provide important information about the dissolution processes of the compounds under study. The apparent standard dissolution enthalpy (ΔH_{sol}^{o}) in the selected solvents can be determined from the van't Hoff analysis (Xu et al. 2018):

$$\Delta H_{sol}^{o} = -R\left(\frac{\partial \ln x}{\partial (1/T - 1/T_{hm})}\right) = -R \cdot slope \qquad (7)$$

where R stands for the universal gas constant. T_{hm} denotes the mean harmonic temperature, which can be calculated by Eq. (8) as 301.47 K:

$$T_{hm} = \frac{n}{\sum_{i=1}^{n}\frac{1}{T_i}} \qquad (8)$$

where N refers to the number of experimental temperature points.

The apparent standard Gibbs energy (ΔG_{sol}^{o}) and entropy (ΔS_{sol}^{o}) were obtained from Eqs. (9) and (10), respectively (Atkins and De Paula, 2006):

$$\Delta G_{sol}^{0} = -RT_{hm} \cdot intercept \tag{9}$$

$$\Delta S_{sol}^{0} = \frac{\Delta H_{sol}^{0} - \Delta G_{sol}^{o}}{T_{hm}} \tag{10}$$

The contributions of enthalpy and entropy (ζ_H and ζ_{TS}) to the standard Gibbs energy in the solution process can be calculated by Eq. (11, 12):

$$\zeta_H = (\Delta H_{sol}^{0}/(\Delta H_{sol}^{0} + T\Delta S_{sol}^{0}))100\% \tag{11}$$

$$\zeta_{TS} = (T\Delta S_{sol}^{0}/(\Delta H_{sol}^{0} + T\Delta S_{sol}^{0}))100\% \tag{12}$$

Partition Experiment

The experiments for determination of the partition coefficients in 1-octanol/buffer (pH 2.0 and 7.4) systems were carried out by the isothermal saturation method at five temperatures: 293.15, 298.15, 303.15, 308.15 and 313.15 K. Before conducting the experiment, both solvents were mutually saturated by equilibrating them in a sufficiently large vessel. This was achieved by slowly stirring them in a biphasic system for two days. The procedure was as follows: a given volume of 1-octanol saturated buffer of a given compound concentration was added to an identical volume of a buffer saturated octanol solution in an ampoule placed into a thermostat. The resulting solution was equilibrated for three days under continuous shaking. The drug concentrations in both phases were determined by a Cary-50 spectrophotometer (USA) in the UV spectral region ($\lambda = 190 - 400$ nm) with an accuracy of 2 – 4%. The reported experimental values represent the average of at least three replicated experiments.

The 1-octanol/aqueous buffer partition coefficients $P_{o/b}$ were calculated as the ratio of the equilibrium concentrations in the organic and aqueous phases:

$$P_{o/b} = s_o/s_b \tag{13}$$

where s_o and s_b are the molar concentrations of the compound in the octanol and buffer phases, respectively. Using equation (1), the partition coefficients $P^*_{o/b}$ were calculated as the ratio of the mole fractions of the compound in the octanol (x_o) and buffer (x_b) phases:

$$P^*_{o/b} = x_o/x_b \qquad (14)$$

The standard Gibbs energy of transfer $\Delta_{tr}G°$ from the buffer to the organic phase was calculated as follows:

$$\Delta_{tr}G° = -RT\ln P^*_{O/B} \qquad (15)$$

The temperature dependence of partition (van't Hoff method) was employed to obtain the enthalpy of transfer $\Delta_{tr}H°$:

$$\frac{d(\ln P^*_{O/B})}{dT} = \frac{\Delta_{tr}H°}{RT^2} \qquad (16)$$

The entropy of transfer $\Delta_{tr}S°$ can be calculated from

$$\Delta_{tr}S° = (\Delta_{tr}H° - \Delta_{tr}G°)/T \qquad (17)$$

Results and Discussion

Chemistry

The synthesis of the hybrids of thiazolo[4,5-d]pyrimidines with (1H-1,2,4)triazole (Groups 1 and 2) was carried out according to Figure 1. Triazole oxirane was synthesized by the known method from 2,4-difluoroacetophenone by treating it with trimethylsulfoxonium iodide in toluene (Fromtling, Castaer, and Serradell, 1985). The sulfur-containing fragment in the thiazolo[4,5-d]pyrimidines was prepared by the known method developed by Gewald, (1966) and described for one of the compounds with a fluorine substituent in the para-position of the phenyl core in position 3 of the thiazole core (Fahmy et al., 2003).

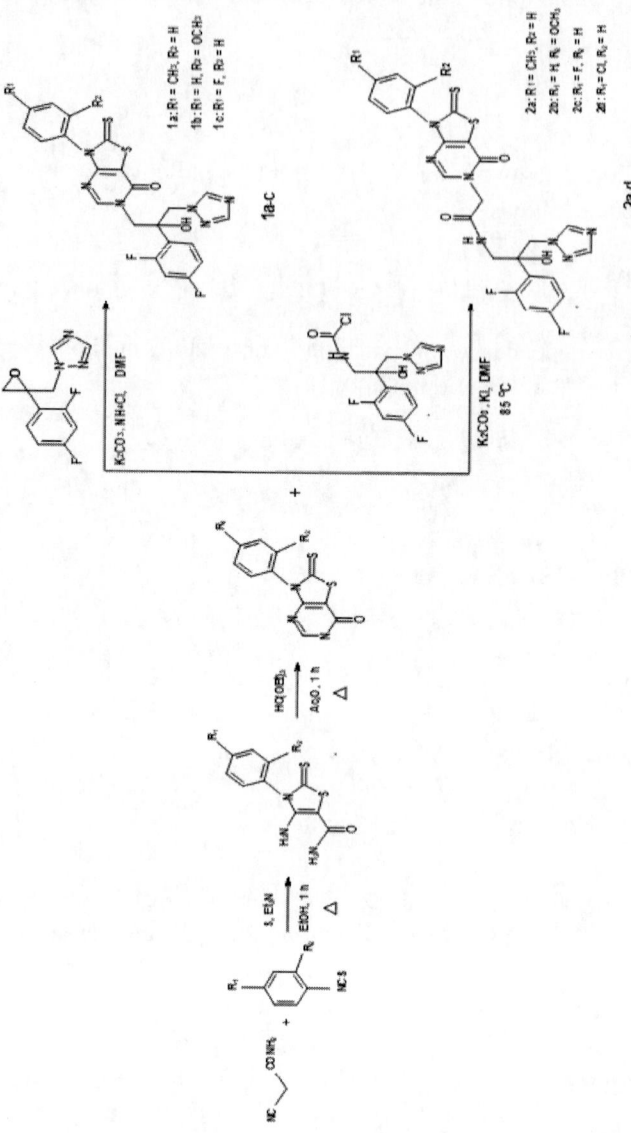

Figure 1. Synthesis of the hybrids of thiazolo[4,5-d]pyrimidines with (1H-1,2,4)triazole (compounds of Groups **1** and **2**).

As the first step, we studied and tested the scheme of synthesizing thiazolo-pyrimidine hybrids with triazole through direct condensation of the derivatives of dihydrothiazolopyrimidine with triazole oxirane prepared in advance. Among the possible variants of preparation of the target products in different solvents – acetonitril, ethanol, toluene with triethylamine and ammonium chloride additives – we selected synthesis with triazole oxirane in a dimethyl formamide for 2-3 days with the formation of 6-[2-(2,4-difluorophenyl)-2-hydroxy-3-(1H-1,2,4-triazol-1-yl)propyl]-3-(R1,R2-phenyl)-2-thioxo-thiazolo[4,5-d]pyrimidin-7-one (**1a-c**) that had not been described earlier. Three compounds with different substituents in the phenyl ring were obtained: R_1 – CH_3, R_2 – H (**1a**); R_1 – H, R_2 – OCH_3 (**1b**) and R_1 – F, R_2 – H (**1c**), their molecular structures are shown in Figure 1.

The reaction of the acetamide triazole chloro-derivative (prepared from triazole oxirane in the process of consecutive interaction with aqueous ammonia and then with chloroacetyl chloride in methylene chloride) with thiazolo pyrimidines leads to the formation of hybrid derivatives with an amide linker (Group 2). The target 6-[2-(2,4 - difluorophenyl) – 2 - hydroxy- 3 -(1H-1,2,4-triazol-1-yl)propyl]-2-[3-(R1,R2-phenyl)-7-oxo-2-thioxo-thiazolo[4,5-d]pyrimidin-6-yl]acetamides (**2a-d**) were obtained.

Antifungal Activity

The serial dilution micromethod was used to determine the antifungal activity of the substances under consideration. The antifungal activity was studied on the following cultures: *C. parapsilosis ATCC 22019, C. albicans ATCC 24433, C. albicans 8R, C. albicans CBS8837, C. albicans 604M, C. utilis84, C. tropicalis 3019, C. glabrata 61 L, C. krusei 432M, Cryptococcus neoformans, A.niger 37a yeast fungi* and *filamentous fungi: M. canis B-200* and *T. rubrum 2002*. Commercially available antifungal agents such as Fluconazole were used as positive control drugs for comparison. The results of the study of the MIC values of the hybrid compounds are summarized in Table 1.

As shown, by combining triazole and thiazolo-pyrimidine fragments in a hybrid molecule we managed to obtain compounds with a wide spectrum of antifungal activity. All the synthesized compounds exhibited microbiological activity against the used strains comparable to a known drug. The results of the biological tests showed that the change in the bridge group structure and introduction of substituents of different chemical nature into the phenyl ring does not have a significant effect on the substance antifungal activity. The most active substance among those studied is derivative **1c**.

Table 1. MIC results of the hybrid compounds studied

Compound	Test microorganisms			
	C. parapsilosis ATCC 22019	C. albicans ATCC 24433	A. niger 37a	M. canis B-200
1a	32	32	32	32
1b	32	32	32	32
1c	8	32	32	32
2a	32	32	32	32
2b	32	32	32	32
2c	32	32	32	32
2d	32	32	32	32
Fluconazole	2	32	64	32

*The averaged data of 3 repeated experiments are given.

Solid State Characterization

The study of the thermophysical properties of the synthesized thiazolo[4,5-d]pyrimidine derivatives by the DSC method has shown that there are no thermal effects when the samples are heated within the temperature range of 20 - 200°C. When the temperature increases to 200 °C, the studied compounds remain stable and do not decompose. As an example, Figure 2a shows the DSC curve of methyl-substituted compound **1a**. The obtained data show that the studied compounds are amorphous. We used the PXRD method to confirm the amorphous state of the synthesized compounds. The absence of reflections on the X-ray patterns confirms that they are amorphous and agrees with the DSC results (Figure 2b). The diffraction patterns indicate that the synthesized derivatives remain amorphous for six months at the temperature of 4°C.

The amorphous state of the synthesized triazole derivatives is associated with the structure of the substance molecules that contain one strong hydrogen bond donor (hydroxyl group, -OH) and 11 potential strong hydrogen bond acceptors (six N-, two F-, one O= and two S= atoms). They have three rotatable groups, namely 2,4-difluorophenyl-, 1,2,4-triazole and R_1, R_2-phenyl. Thus, the available options for molecular association in the condensed phase are limited only to weak intermolecular interactions of the C-H...N, C-H...O, C-H...F, C-H...π-π, etc. types. The absence of strong hydrogen bond donor groups and high acceptor/donor ratio are the causes of the formation of a comparatively low-ordered amorphous phase by the compounds studied.

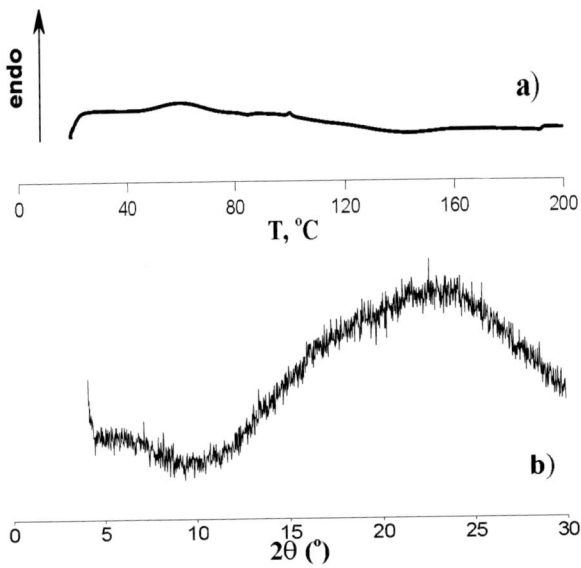

Figure 2. Typical DSC curve (a) and diffraction pattern (b) for the compounds studied (1a as an example).

Study of Dissolution and Transfer Processes of the Hybrid Compounds of Group I

Kinetic and Equilibrium Solubility
The kinetic solubility curves of the synthesized thiazolo[4,5-d]pyrimidine derivatives - **1a-c** – in buffer solutions pH 2.0 and 7.4 are shown in Figure 3. The obtained results allow us to conclude that the structural changes in the substances caused by the introduction of substituents into the phenyl ring affect the kinetic solubility characteristics. In buffer pH 2.0, the kinetic solubility of the amorphous samples of methyl- and methoxy- substituted compounds **1a** and **1b** is higher than in buffer pH 7.4. In contrast, the changes in the acidity of the aqueous solution do not have a strong effect on the kinetic solubility of fluoroderivative **1c**. The dependences of the concentration increase for compounds **1a** and **1c** in aqueous solutions as well as for compound **1b** in buffer pH 7.4 on time are similar. The kinetic curves in these systems have an initial region with a fast increase in the substance content in the solution and then a region with a slower increase in the concentration until it reaches a constant value.

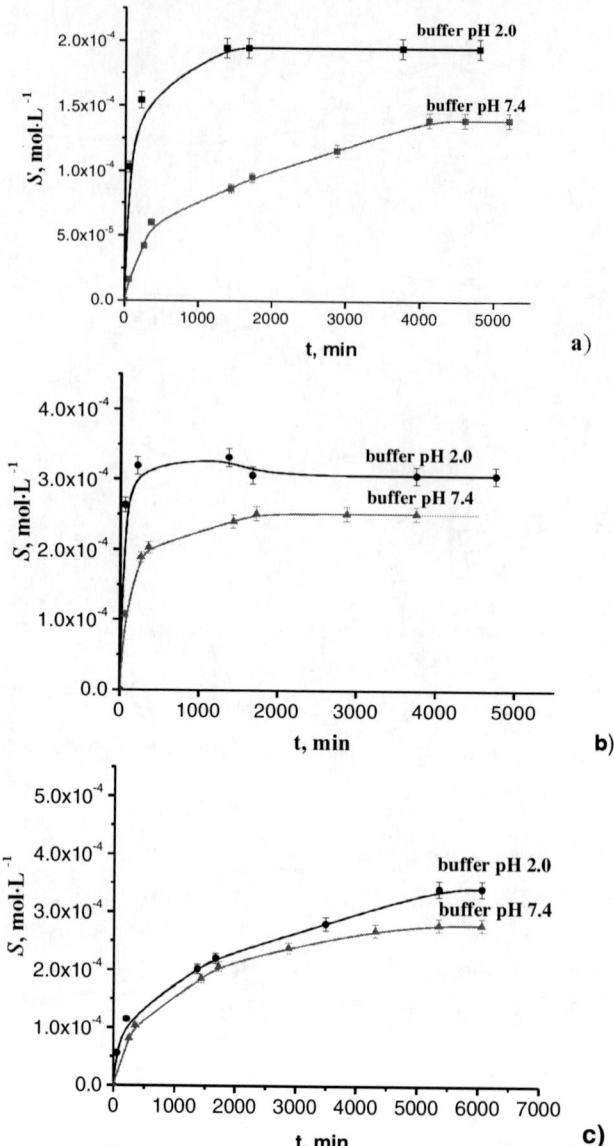

Figure 3. Kinetic solubility curves of compounds **1a-c** in buffers pH 2.0 and 7.4 at 298.15 K.

The solubility curve of compound **1b** in buffer pH 2.0 has a supersaturated solution region that is transformed into a plateau with a lower kinetic solubility value. The experiments made show that the time required to reach equilibrium in the studied systems changes from two to four days. The study of the bottom phases by the DSC method conducted after the experiments did not show any changes in the amorphous state of the substances.

The solubility of thiazolo[4,5-d]pyrimidine derivatives **1(a - c)** was determined in pharmaceutically relevant solvents: buffer pH 2.0, buffer pH 7.4 and 1-octanol in the temperature range (293.15 – 313.15) K by the shake-flask method (Table 2). The experimental data are presented graphically in Figure 4.

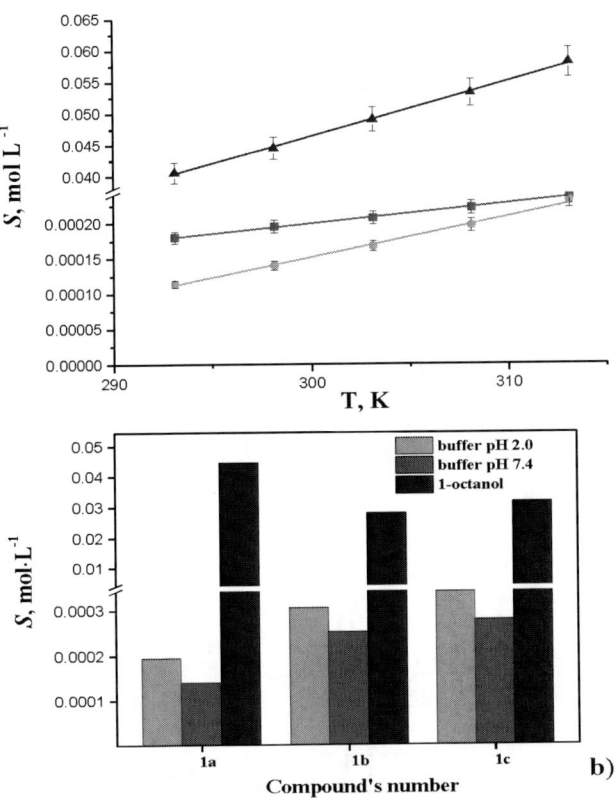

Figure 4. Solubility data of the compounds studied in the selected solvents: a) Temperature dependence of equilibrium solubility for compound **1a** in the solvents studied: ● – buffer pH 2.0, ■ – buffer pH 7.4, ▲ –1-octanol; b) solubility histogram at 298.15 K.

Table 2. Temperature dependences of solubility (S, mol·L^{-1}) of the compounds studied (amorphous state) in buffer solutions (pH 2.0[a] and 7.4[b]) and 1-octanol at pressure p=0.1 MPa

T/K	1a			1b			1c		
	buffer pH 2.0	buffer pH 7.4	1-octanol	buffer pH 2.0	buffer pH 7.4	1-octanol	buffer pH 2.0	buffer pH 7.4	1-octanol
	S·10^4	S·10^4	S·10^2	S·10^4	S·10^4	S·10^2	S·10^4	S·10^4	S·10^2
293.15	1.80	1.14	4.06	2.69	2.07	2.48	3.21	2.38	2.92
298.15	1.95	1.40	4.45	3.06	2.52	2.79	3.42	2.79	3.16
303.15	2.07	1.67	4.90	3.39	3.09	3.06	3.59	3.31	3.44
308.15	2.21	1.96	5.32	3.85	3.79	3.41	3.81	3.91	3.68
313.15	2.37	2.31	5.80	4.29	4.65	3.71	4.01	4.52	4.00

[a]Composition of aqueous buffer pH 2.0: KCl (6.57 g in 1 L) and 0.1 mol·L^{-1} hydrochloric acid (119.0 mL in 1 L);
[b]Composition of aqueous buffer pH 7.4: KH$_2$PO$_4$ (9.1 g in 1 L) and Na$_2$HPO$_4$·12H$_2$O (23.6 g in 1 L).
Standard uncertainties for the salt mass (m) and solution volume (V): $u(m)$ = 5 mg, $u(V)$ = 0.5 mL.
The standard uncertainties are $u(T)$ = 0.15 K, $u(p)$ = 3 kPa, u(pH) = 0.02 pH units.
The relative standard uncertainties are $u_r(x)$ = 0.04 and $u_r(S)$ = 0.04.

The analysis of the obtained data shows that the solubility of the substances grows as the temperature increases. The solubility values vary within the range (2-5) 10^{-4} mol·L^{-1} in the aqueous solvents and (2 - 6) 10^{-2} mol·L^{-1} in 1-octanol. The increase in the solution acidity from buffer pH 7.4 to buffer pH 2.0 makes the solubility of the compounds higher, which is associated with the transition of the neutral form of molecules to the ionized state. The compounds studied have proton-acceptor atoms in their structure, and in the acidic medium of buffer pH 2.0 can exist in the form of cations.

The pKa values calculated in the program *p*DISOL-X (Avdeef, 2014) corresponding to the protonation of two nitrogen atoms of 1,2,4-triazole are approximately equal to 2.3, while the pKa value corresponding to the protonation of the nitrogen atom of thiazolo[4,5-d]pyrimidine is 2.6. This fact is explained by the basic nature of 1,2,4-triazole and heterocycles with nitrogen atoms in the structure of the synthesized substances. In comparison, fluconazol is also protonated at two nitrogen atoms of 1,2,4-triazole and has pKa values: 2.94 ± 0.10 and 2.56 ± 0.12 (Correa et al., 2012). The introduction of the considered electron-donor substituents has little effect on the protolytic properties of the synthesized derivatives. It was found that in the ascending order of solubility values in the studied aqueous solvents, the compounds can be arranged as follows: **1a** (-CH$_3$) <**1b** (-OCH$_3$) <**1c** (-F). It can be assumed that the shift in the benzene ring electron density towards the electronegative atom leads to the emergence of a bond dipole moment that increases in the series: methyl-, methoxy- and fluoro-. In turn, the increase in the polarity of the molecules of the synthesized compounds makes the dipole-dipole interaction with the solvent molecules stronger and, as a result, the solubility of the substances grows.

The solubility of the studied thiazolo[4,5-d]pyrimidine derivatives in 1-octanol is approximately two orders of magnitude higher than in the aqueous solvents. The data on the solubility of the known antifungal drugs of the azole class show that their solubility in alcohols mainly depends on the ability to form hydrogen bonds with the solvent molecules (Xie et al., 2016). Therefore, it can be concluded that the comparatively high solubility of the compounds in 1-octanol is caused by stronger specific interactions of the dissolved substances with the alcohol molecules than with the water ones.

Amorphization is one of the most common ways of obtaining soluble forms of drug compounds. The amorphous form of the obtained thiazolo[4,5-d]pyrimidine derivatives largely determines their physico-chemical properties, including solubility. And it should be taken into account that better solubility can be achieved through the compound transition from the stable crystalline state to the metastable amorphous one. During the solubility determination procedure by the shake flask method, the compound is redeposited a lot of times, which can have an effect on its state. This fact made us study the bottom phases by the DSC method. The absence of thermal effects on the curves obtained (Figure 2) allowed us to classify the solid state of the studied thiazolo[4,5-d]pyrimidine derivatives as amorphous. The obtained results agree with the studies of the substance residues after the kinetic solubility determination. Besides, the amorphous form sustainability of the substances is confirmed by the fact that the kinetic solubility value coincides with that of the thermodynamic solubility in stable equilibrium conditions.

Modeling of Solubility Data

In order to more quantitatively describe the solid–liquid equilibrium, the relationship between the experimental solubility and the temperature was modeled by the modified Apelblat and van't Hoff equations. The experimental values of the compound solubility and those calculated by Eq. (1-3) are presented in Table 3. The parameters of the modified Apelblat and van't Hoff equations for the compounds studied in the selected solvents are presented in Table 4.

The *RMSD, RAD* and *RD values* calculated by Eq. (4 - 6) are shown in Tables 3 and 4. The results suggest that the correlated values of the compounds in the solvents are in accordance with the experimental solubility data.

Table 3. Experimental (x_{exp}) and correlated (x_{cal}) mole fractions of solubility of the compounds studied in the specified solvents at different temperatures

T/K	x_{exp}	Modified Apelblat equation		van't Hoff equation	
		x_{cal}	100RD	x_{cal}	100RD
1a					
buffer pH 2.0					
293.15	$3.25 \cdot 10^{-6}$	$3.26 \cdot 10^{-6}$	-0.20	$3.25 \cdot 10^{-6}$	-0.04
298.15	$3.51 \cdot 10^{-6}$	$3.50 \cdot 10^{-6}$	0.39	$3.50 \cdot 10^{-6}$	0.36
303.15	$3.74 \cdot 10^{-6}$	$3.75 \cdot 10^{-6}$	-0.23	$3.75 \cdot 10^{-6}$	-0.31
308.15	$4.01 \cdot 10^{-6}$	$4.02 \cdot 10^{-6}$	-0.19	$4.02 \cdot 10^{-6}$	-0.20
313.15	$4.30 \cdot 10^{-6}$	$4.30 \cdot 10^{-6}$	0.09	$4.29 \cdot 10^{-6}$	0.25
buffer pH 7.4					
293.15	$2.08 \cdot 10^{-6}$	$2.09 \cdot 10^{-6}$	-0.48	$2.10 \cdot 10^{-6}$	-0.83
298.15	$2.54 \cdot 10^{-6}$	$2.54 \cdot 10^{-6}$	0.12	$2.53 \cdot 10^{-6}$	0.70
303.15	$3.04 \cdot 10^{-6}$	$3.05 \cdot 10^{-6}$	-0.12	$3.02 \cdot 10^{-6}$	0.74
308.15	$3.58 \cdot 10^{-6}$	$3.61 \cdot 10^{-6}$	-0.88	$3.59 \cdot 10^{-6}$	-0.34
313.15	$4.23 \cdot 10^{-6}$	$4.23 \cdot 10^{-6}$	0.02	$4.25 \cdot 10^{-6}$	-0.34

T/K	x_{exp}	Modified Apelblat equation		van't Hoff equation	
		x_{cal}	100RD	x_{cal}	100RD
1a					
1-octanol					
293.15	6.51·10⁻³	6.51·10⁻³	0.05	6.51·10⁻³	0.01
298.15	7.19·10⁻³	7.20·10⁻³	-0.23	7.20·10⁻³	-0.17
303.15	7.97·10⁻³	7.94·10⁻³	0.29	7.94·10⁻³	0.37
308.15	8.71·10⁻³	8.72·10⁻³	-0.20	8.72·10⁻³	-0.15
313.15	9.55·10⁻³	9.55·10⁻³	0.03	9.55·10⁻³	-0.02
1b					
buffer pH 2.0					
293.15	4.86·10⁻⁶	4.87·10⁻⁶	-0.22	4.86·10⁻⁶	-0.01
298.15	5.53·10⁻⁶	5.49·10⁻⁶	0.66	5.49·10⁻⁶	0.59
303.15	6.14·10⁻⁶	6.18·10⁻⁶	-0.77	6.19·10⁻⁶	-0.92
308.15	6.97·10⁻⁶	6.95·10⁻⁶	0.33	6.95·10⁻⁶	0.27
313.15	7.79·10⁻⁶	7.79·10⁻⁶	-0.05	7.77·10⁻⁶	0.16
buffer pH 7.4					
293.15	3.72·10⁻⁶	3.70·10⁻⁶	0.50	3.68·10⁻⁶	1.07
298.15	4.52·10⁻⁶	4.51·10⁻⁶	0.15	4.56·10⁻⁶	-0.79
303.15	5.56·10⁻⁶	5.53·10⁻⁶	0.52	5.61·10⁻⁶	-0.89
308.15	6.83·10⁻⁶	6.80·10⁻⁶	0.50	6.85·10⁻⁶	-0.36
313.15	8.40·10⁻⁶	8.37·10⁻⁶	0.33	8.32·10⁻⁶	0.91
1-octanol					
293.15	3.95·10⁻³	3.94·10⁻³	0.44	3.93·10⁻³	0.52
298.15	4.47·10⁻³	4.43·10⁻³	1.03	4.40·10⁻³	1.62
303.15	4.94·10⁻³	4.94·10⁻³	-0.04	4.90·10⁻³	0.71
308.15	5.53·10⁻³	5.47·10⁻³	0.93	5.44·10⁻³	1.49
313.15	6.06·10⁻³	6.03·10⁻³	0.50	6.02·10⁻³	0.56
1c					
buffer pH 2.0					
293.15	5.79·10⁻⁶	5.80·10⁻⁶	-0.13	5.79·10⁻⁶	0.07
298.15	6.17·10⁻⁶	6.14·10⁻⁶	0.37	6.15·10⁻⁶	0.22
303.15	6.50·10⁻⁶	6.51·10⁻⁶	-0.21	6.53·10⁻⁶	-0.48
308.15	6.90·10⁻⁶	6.91·10⁻⁶	-0.05	6.92·10⁻⁶	-0.19
313.15	7.33·10⁻⁶	7.32·10⁻⁶	0.09	7.31·10⁻⁶	0.29
buffer pH 7.4					
293.15	4.27·10⁻⁶	4.25·10⁻⁶	0.35	4.25·10⁻⁶	0.47
298.15	5.00·10⁻⁶	5.04·10⁻⁶	-0.70	5.04·10⁻⁶	-0.79
303.15	5.95·10⁻⁶	5.95·10⁻⁶	0.12	5.96·10⁻⁶	-0.04
308.15	7.04·10⁻⁶	6.99·10⁻⁶	0.62	7.00·10⁻⁶	0.54
313.15	8.16·10⁻⁶	8.19·10⁻⁶	-0.31	8.18·10⁻⁶	-0.19
1-octanol					
293.15	4.66·10⁻³	4.66·10⁻³	-0.08	4.66·10⁻³	0.07
298.15	5.08·10⁻³	5.08·10⁻³	0.04	5.08·10⁻³	-0.01
303.15	5.55·10⁻³	5.53·10⁻³	0.37	5.53·10⁻³	0.26
308.15	5.97·10⁻³	6.01·10⁻³	-0.55	6.01·10⁻³	-0.59
313.15	6.53·10⁻³	6.52·10⁻³	0.21	6.51·10⁻³	0.37

Standard uncertainities: $u(T) = 0.15$ K and $u(p) = 3$ kPa.
Relative standard uncertainties for solubility: $u_r(x) = 0.045$ for buffer solutions and $u_r(x) = 0.04$ for 1-octanol.

Table 4. Parameters of the modified Apelblat and van't Hoff equations for the compounds studied in the selected solvents

Solvents	A	B	C	RMSD	100RAD
1a					
Modified Apelblat equation					
buffer pH 2.0	-39.11	122.6	4.59	$8.64 \cdot 10^{-9}$	0.22
buffer pH 7.4	149.11	-10055.9	-22.54	$1.25 \cdot 10^{-8}$	0.31
1-octanol	14.85	-2387.4	-2.07	$1.49 \cdot 10^{-6}$	0.31
van't Hoff equation					
buffer pH 2.0	-8.31	-1266.8	-	$9.74 \cdot 10^{-9}$	0.23
buffer pH 7.4	-2.04	-3235.0	-	$1.71 \cdot 10^{-8}$	0.59
1-octanol	0.97	-1761.2	-	$1.54 \cdot 10^{-5}$	0.14
1b					
Modified Apelblat equation					
buffer pH 2.0	-49.42	-149.5	6.64	$2.89 \cdot 10^{-8}$	0.41
buffer pH 7.4	-246.71	7394.2	36.79	$2.49 \cdot 10^{-8}$	0.40
1-octanol	86.30	-5797.9	-12.68	$3.46 \cdot 10^{-5}$	0.59
van't Hoff equation					
buffer pH 2.0	-4.87	-2159.8	-	$3.08 \cdot 10^{-8}$	0.39
buffer pH 7.4	-0.28	-3749.3	-	$4.86 \cdot 10^{-8}$	0.80
1-octanol	1.14	-1955.4	-	$5.45 \cdot 10^{-5}$	0.98
1c					
Modified Apelblat equation					
buffer pH 2.0	-66.15	-1533.9	8.60	$4.78 \cdot 10^{-9}$	0.17
buffer pH 7.4	-37.17	-1427.7	5.22	$2.84 \cdot 10^{-8}$	0.42
1-octanol	-32.44	-78.5	4.81	$1.85 \cdot 10^{-5}$	0.25
van't Hoff equation					
buffer pH 2.0	-8.40	-1071.7	-	$1.88 \cdot 10^{-8}$	0.25
buffer pH 7.4	-2.10	-3009.9	-	$2.70 \cdot 10^{-8}$	0.41
1-octanol	-0.13	-1536.2	-	$2.01 \cdot 10^{-5}$	0.26

In this study, the van't Hoff and modified Apelblat models were used to correlate the temperature dependences of the solubility of the compounds studied in different solvents. It can be seen that these thermodynamic models evalute the solubility data with *RAD* values of less than 1%. The mean values of 100RD calculated based on the data of Table 3 show changes in this parameter for the studied solute-solvent systems. The relative deviations increase in the following order: buffer pH 2.0 (0.26) < buffer pH 7.4 (0.38) < 1-octanol (0.41) for the modified Apelblat model, while for the van't Hoff model they can be arranged in the following series: buffer pH 2.0 (0.29) < 1-octanol (0.54) < buffer pH 7.4 (0.60). An analysis of the data obtained indicates that both thermodynamic models provide more accurate correlation results for the solubility values of the compounds in buffer pH 2.0 than in the other solvents. Moreover, the minimal values of average 100RD are observed for the Apelblat model, demonstrating that this equation describes the experimental values better than the van't Hoff model. According to the values of relative root-mean square deviation ($RMSD$) (Table 4), the Apelblat model is well fitted to the measured solubility data of the compounds studied in the selected solvents with the relative root mean square deviation changing from 0.02 to 1.37%.

Apparent Thermodynamic Functions of Solution

Apparent thermodynamic analysis is widely used to evaluate the dissolution behavior of a solute (Delgado et al., 2021). The molar dissolution thermodynamic properties of the compounds were calculated and analyzed based on the experimental solubility data. Figure 5 shows the modified van't Hoff plot for the compounds in the selected solvents. In all the cases, linear models with good determination coefficients were obtained. The parameters of the equation $\ln x = A - B(1/T - 1/T_{hm})$ are given in Table 5. The apparent thermodynamic functions of dissolution of compound **1a-c** in the studied solvents obtained by equations (7-10) based on the solubility temperature dependences are shown in Table 6. The obtained values of the Gibbs energy, enthalpy and entropy can be used to compare the substance dissolution in aqueous and organic solvents.

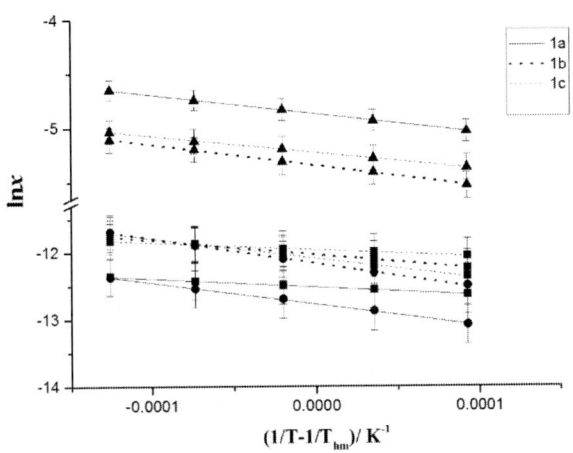

Figure 5. Temperature dependence of $\ln x$ for compounds **1a - c** in the solvents studied: ■– buffer pH 2.0, ●– buffer pH 7.4, ▲–1-octanol.

Table 5. Coefficients of the correlation equation $\ln x = A - B/(1/T - 1/T_{hm})$

Compound	A	B	R
buffer pH 2.0			
1a	-12.52 ± 0.01	1265 ± 19	0.9996
1b	-12.03 ± 0.01	2158 ± 37	0.9995
1c	-11.96 ± 0.01	1070 ± 21	0.9995
buffer pH 7.4			
1a	-12.77 ± 0.01	3231 ± 46	0.9997
1b	-12.17 ± 0.01	3745 ± 62	0.9996
1c	-12.09 ± 0.01	3007 ± 36	0.9998
1-octanol			
1a	-4.87 ± 0.01	1759 ± 15	0.9999
1b	-5.35 ± 0.01	1953 ± 35	0.9994
1c	-5.02 ± 0.01	1534 ± 24	0.9996

As Table 6 shows, all of the computed values of the apparent standard dissolution enthalpy are positive suggesting that the compounds dissolution processes are endothermic in all of the solvents. The weaker intermolecular force between the solvent molecules than between the solute molecules may cause this phenomenon. In detail, the newly formed bond energy between the compound and solvent molecules are not enough to break the original associative bonds of the solvent. The values of apparent molar standard Gibbs energy are positive, which indicates difficulty of solution formation. The highest and the lowest values of ΔG_{sol}^o are observed in buffer pH 7.4 and 1-octanol, respectively. The apparent standard entropies of dissolution in buffer solutions pH 2.0 and pH 7.4 are negative and the dissolution process is enthalpy-driven. On the contrary, the values of ΔS_{sol}^o in 1-octanol are positive and the dissolution process is entropy-driven. Exceptions are the following systems: compound **1b** in buffer pH 7.4 and compound **1c** in 1-octanol, with weakly positive and weakly negative values of entropy, respectively. The negative dissolution entropy observed in buffer solutions can be explained in terms of the possible hydrophobic hydration around the non-polar groups of the compounds studied (Jouyban, Acree, and Martínez, 2020).

As Table 6 also shows, the main contributor to the positive standard molar Gibbs energy of dissolution of the compounds in buffer solute pH 7.4 and octanol is the positive enthalpy ($\zeta_H > 80$), which energetically prevails over all the dissolution processes.

Table 6. Apparent thermodynamic solubility functions of the compounds studied in buffer pH 2.0, buffer pH 7.4 and 1-octanol at T_{im}=301.47 K

Compound	ΔG^o_{sol} (kJ/mol)	ΔH^o_{sol} (kJ/mol)	$T\Delta S^o_{sol}$ (kJ/mol)	ΔS^o_{sol} (J/mol·K)	ζ_H, %	ζ_{TS}, %
	buffer pH 2.0					
1a	31.4± 0.4	10.5 ± 0.2	-20.9	-69.3± 1.4	33.4	66.6
1b	30.2 ± 0.4	17.9 ± 0.3	-12.3	-40.8 ±0.3	59.3	40.7
1c	30.0 ± 0.4	8.9 ± 0.1	-21.1	-70.0± 0.8	29.7	70.3
	buffer pH 7.4					
1a	32.0 ± 0.4	26.9 ± 0.4	-5.1	-16.9± 0.3	84.1	15.9
1b	30.5 ± 0.4	31.1 ± 0.5	0.6	2.1 ±1.2	98.1	1.9
1c	30.3 ± 0.4	25.0 ± 0.3	-5.3	-17.6 ±0.6	82.5	17.5
	1-octanol					
1a	12.2 ± 0.1	14.6 ± 0.1	2.4	8.0± 0.5	85.9	14.1
1b	13.4 ± 0.2	16.2 ± 0.3	2.8	9.3 ±0.8	85.3	14.7
1c	13.1± 0.2	12.8 ± 0.2	-0.3	-1.0 ± 0.4	98.5	1.5

Lipophilicity

When screening new potential drug compounds, it is important to study their lipophilicity that affects passive transport of substances through biological membranes. The lipophilicity was estimated by measuring the partition coefficients of the neutral forms of the synthesized substances in the 1-octanol/buffer pH 7.4 system in the temperature range (293.15 – 313.15) K. The partition coefficients were determined as the ratio of the molarity concentrations of the substances in the organic phase to those in the aqueous phase (eq. 13, 14). They are summarized in Table 7, where the parameters of the correlation equation $\log P^*_{o/b} = A + B/T$ and the paired correlation coefficient (R) are also indicated.

In the studied temperature range, the $\log P_{o/b}$ values range from 0.43 to 1.44, which indicates more favourable distribution of the thiazolo[4,5-d]pyrimidine derivatives in the organic phase (a hydrophobic environment) than in the aqueous phase (a hydrophilic environment). As the temperature rises, the equilibrium shifts to 1-octanol and the partition coefficients increase, which results from higher lipophilicity of the compounds.

Table 7. Experimental concentrations (s_o, s_b in mol·L^{-1} and x_o, x_b in mole fraction in 1-octanol and buffer, respectively) and partition coefficients ($\log P_{o/b}$, $\log P^*_{o/b}$ are calculated based on concentrations expressed in mol·L^{-1} and mole fraction, respectively) for the compounds studied in the 1-octanol/buffer pH 7.4 system at different temperatures

T/K	1a						1b					
	$s_o \cdot 10^4$	$s_b \cdot 10^5$	$\log P_{o/b}$	$x_o \cdot 10^5$	$x_b \cdot 10^7$	$\log P^*_{o/b}$	$s_o \cdot 10^4$	$s_b \cdot 10^6$	$\log P_{o/b}$	$x_o \cdot 10^5$	$x_b \cdot 10^7$	$\log P^*_{o/b}$
293.15	1.59	2.66	0.78	2.51	4.76	1.72	1.20	9.11	1.12	1.89	2.66	1.85
298.15	1.60	2.59	0.79	2.53	4.64	1.74	1.21	7.75	1.19	1.92	2.43	1.90
303.15	1.61	2.53	0.80	2.55	4.55	1.75	1.22	6.49	1.28	1.95	2.22	1.94
308.15	1.61	2.48	0.81	2.57	4.46	1.76	1.23	5.45	1.36	1.97	2.04	1.98
313.15	1.62	2.43	0.82	2.59	4.37	1.77	1.24	4.50	1.44	1.99	1.89	2.02
A^a	5.77 ± 0.02						10.50 ± 0.05					
B^a	529 ± 5						1828 ± 15					
R^b	0.9999						0.9998					

T/K	1c					
	$s_o \cdot 10^5$	$s_b \cdot 10^5$	$\log P_{o/b}$	$x_o \cdot 10^6$	$x_b \cdot 10^7$	$\log P^*_{o/b}$
293.15	4.69	1.71	0.43	7.39	3.05	1.38
298.15	4.79	1.61	0.47	7.57	2.88	1.42
303.15	4.87	1.52	0.50	7.74	2.73	1.45
308.15	4.95	1.44	0.53	7.90	2.59	1.48
313.15	5.02	1.37	0.56	8.04	2.47	1.51
A^a	7.81 ± 0.08					
B^a	1354 ± 23					
R^b	0.9996					

It was found that the chemical nature of the substituents introduced into the phenyl ring has a significant effect on the distribution of the studied compounds. In the descending order of partition coefficient values the compounds are arranged as: **1b** ($-OCH_3$) > **1a** ($-CH_3$) > **1c** ($-F$). Bioactive compounds, for which the inequality $1 < \log P_{o/b} < 3$ is satisfied, are known to have a good balance between the solubility and permeability through passive diffusion, which makes them highly absorbable (Kerns and Di, 2008). The $\log P_{o/b}$ values of the studied derivatives at the temperature 298.15 K are equal to 0.79, 1.19 and 0.47 for compounds **1a-c**, respectively. Consequently, compound **1b** with a methoxy-group as the substituent has the optimal lipophilicity for its application in pharmaceutics.

The temperature dependences of the partition coefficients (Figure 6) were used to calculate the transfer thermodynamic functions of the compounds in the 1-octanol/buffer pH 7.4 system at 298.15 K by equations (15-17) (Table 8). The quantities $\Delta_{tr}H^o$ and $\Delta_{tr}S^o$ represent, respectively, the change in enthalpy and entropy when one mole of the solute is transferred from the aqueous phase to the octanol one (Dearden and Bresnen, 2005).

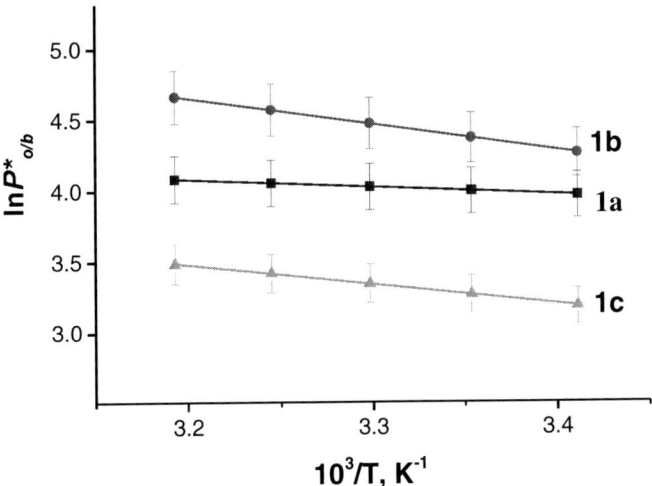

Figure 6. Temperature dependence of the partition coefficients for compounds **1a-c** in the 1-octanol/buffer pH 7.4 system.

The negative values of Gibbs energy indicate the process of transfer from the hydrophilic to the lipophilic environment for the substances is favorable. In this case, the distribution is controlled by the entropy term of the Gibbs energy: $T\Delta trSo > \Delta trHo$. For all the compounds studied, the enthalpy and entropy changes are positive and, therefore, the partition process is endothermic and entropy-driven. The energy absorption is probably caused by the re-solvation of molecules during the transfer from the aqueous phase to the organic one. In addition, a decrease in the ordering of the system shows that the heterocyclic fragments of the studied substances are poorly compatible with the molecular layers of 1-octanol.

Table 8. Thermodynamic functions of transfer in the 1-octanol/buffer pH 7.4 system for the compounds studied at 298.15 K

Compound	$\Delta_{tr}G°$ kJ·mol^{-1}	$\Delta_{tr}H°$ kJ·mol^{-1}	$T\Delta_{tr}S°$ kJ·mol^{-1}	$\Delta_{tr}S°$ J·mol^{-1}·K^{-1}
1a	-9.9± 0.2	4.4 ± 0.1	14.3± 1.0	48.0 ± 0.2
1b	-10.8 ± 0.2	15.2 ± 0.1	26.0 ± 0.8	87.2 ± 0.4
1c	8.1 ± 0.2	11.3 ± 0.2	19.4 ± 0.9	65.0 ± 0.7

The standard uncertainties are $u(T)=0.15$ K, $u(p) = 3$ kPa.

The molecules of the studied compounds, as in most drugs, consist of heterocyclic fragments and aromatic nuclei with substituents. This makes them capable of protonation, hydrogen bonding and specific solvation with polar molecules of water and 1-octanol, which complicates the use of calculation methods. The partition coefficients of the compounds in the 1-octanol/buffer pH 7.4 system were calculated in different computer programs. In the HYBOT software package for bioactive compounds, the calculations were made by the following equation (Raevsky, Grigor'ev, and Trepalin, 1999):

$$\text{Clog}P_3=0.266(\pm0.006)\alpha-1.00(\pm0.05)\Sigma C_a \quad (18)$$

where the independent variables are represented by the descriptors: polarizability (α) and the sum of all hydrogen-bond acceptor factors (ΣC_a) of the molecules of the studied compounds, the values of which are given in Table 9. It should be noted that the partition coefficients of the compounds correlate with the molecule polarizability values. The experimentally obtained values of log$P_{o/b}$ were compared with those calculated in the ACD/Labs

(ClogP$_1$), HYBOT (ClogP$_2$) and HyperChem (ClogP$_3$) programs based on the residual (Res) and root mean squared error (RMSE) values. The calculation results given in Table 9 show that the ClogP$_1$ values are, on average, one logarithmic unit higher than the measured values of logP$_{O/B}$. When we used the HYBOT program, the mean deviation between the logP$_{o/b}$ and (ClogP$_2$) was 0.6 logarithmic units. The best result was obtained in the HyperChem program: the ClogP$_3$ values differed from the logP$_{o/b}$ ones, on average, by 0.5 logarithmic units

Table 9. Partition coefficients of compounds 1a-c in the 1-octanol/buffer pH 7.4 system: experimental logP$_{o/b}$ and predicted ClogP$_1$, ClogP$_2$, ClogP$_3$ at 298.15 K

Compound	logP$_{o/b}$	α	ΣC$_a$	ClogP$_1$	Res$_1$*	ClogP$_2$	Res$_2$*	ClogP$_3$	Res$_3$*
1a	0.79	50.76	13.73	1.86	-1.07	0.76	0.03	1.16	-0.37
1b	1.19	51.03	14.39	1.69	-0.5	0.21	0.98	0.74	0.45
1c	0.47	48.83	13.72	1.75	-1.28	0.24	0.23	1.05	-0.58
RMSE				1.00553		0.58146		0.47455	

ClogP$_1$ calculated using Advanced Chemistry Development (ACD/Labs) Software V10.02
ClogP$_2$ calculated using HYBOT
ClogP$_3$ calculated using HyperChem 9.0 Molecular Mechanics, Method MM+.
*Res$_i$ = logP$_{O/B}$ − ClogP$_i$ (i=1, 2, 3)

$$RMSE = \left| \frac{1}{N} \sum_{i=1}^{N} \left(\log P_{o/b}^{exp} - \log P_{o/b}^{cal} \right) \right|^{1/2}$$

Correlation between Biological Activity and Solubility

The experiments with the pathogenic fungi strains *C. parapsilosis* ATCC 22019 have shown that triazole derivatives **1b** and **1c** with a methoxy- group and a fluorine atom as the substituents in the phenyl ring possess antimycotic activity. Under the action of the mentioned compounds on the test-strain *C. parapsilosis* ATCC 22019 the minimum inhibitory concentration (MIC) after 24 hours of incubation was 8 µg/ml. After 48 hours it was 16 µg/ml and 8 µg/ml for compounds **1b** and **1c**, respectively. At this concentration, the visible growth in the test-culture was suppressed by at least 80% in comparison with the reference. For comparison, the MIC of fluconazole for the *C. Parapsilosis* ATCC 22019 strain is equal to 2.0 - 4.0 µg/ml. However, the MIC value of compound I was over 32 µg/ml. The studied compounds did not have a noticeable effect on the other test-microorganisms. It should be said that the increase in the biological activity of the synthesized triazole derivatives after the introduction of substituents in the order methyl-, methoxy-, fluoro- corresponds to the solubility growth of the substances in the buffer solutions (Figure 7).

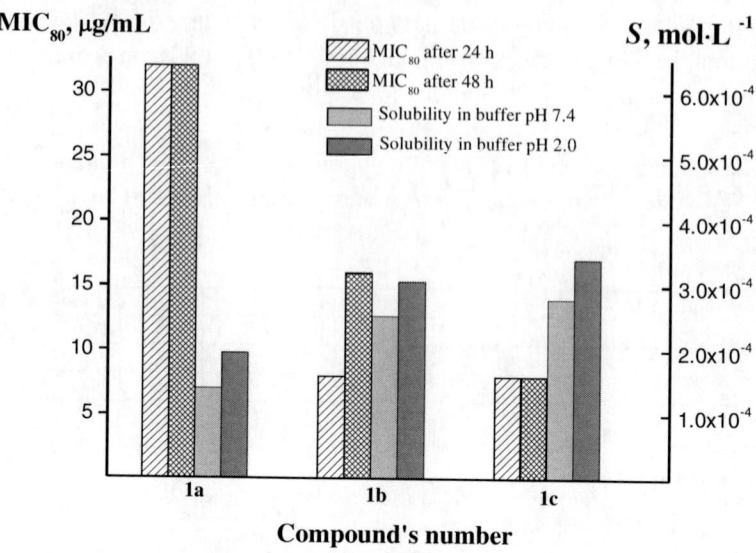

Figure 7. Histogram of MIC values for the strains of pathogenic fungi *C.parapsilosis ATCC 22019* and solubility of compounds **1a–c** in the buffer solutions.

Thermodynamic Insights into Solubility and Lipophilicity of Hybrid Compounds of Group 2

Kinetic Profiles

The kinetic solubility of the synthesized substances was studied in aqueous solutions at physiological values of the gastro-intestinal tract medium acidity (pH 2.0 and 7.4). As the data in Figure 8 show, the dissolution kinetics of the studied derivatives is a two-stage process consisting of a fast increase in the compound concentration in the aqueous solution in the first 250 min of the experiment and then a gradual change in the solubility until a plateau is reached.

The time before the equilibrium solubility value is reached ranges from 2000 to 3500 min depending on the substance structure. In all the studied solute-solvent systems, the solubility in buffer pH 2.0 is higher than in the pH

7.4 one, which is explained by the triazole derivative characteristics. The kinetic solubility curves of the compounds shown in the Figure depend on the chemical nature of the substituents introduced into the molecule phenyl ring. The substance concentration in the solution gradually increases in derivative **2a** with a methyl group, whereas the solubility curve of fluoro-derivative **2d** represents an extremal dependence. The solubility dependences of derivative **2b** with a methoxy-group and derivative **2c** with a chlorine atom on time are practically identical. The X-ray diffraction patterns of the solid phases extracted from the solution after the experiments are completed are absolutely identical to those of the amorphous phases of the initial compound samples.

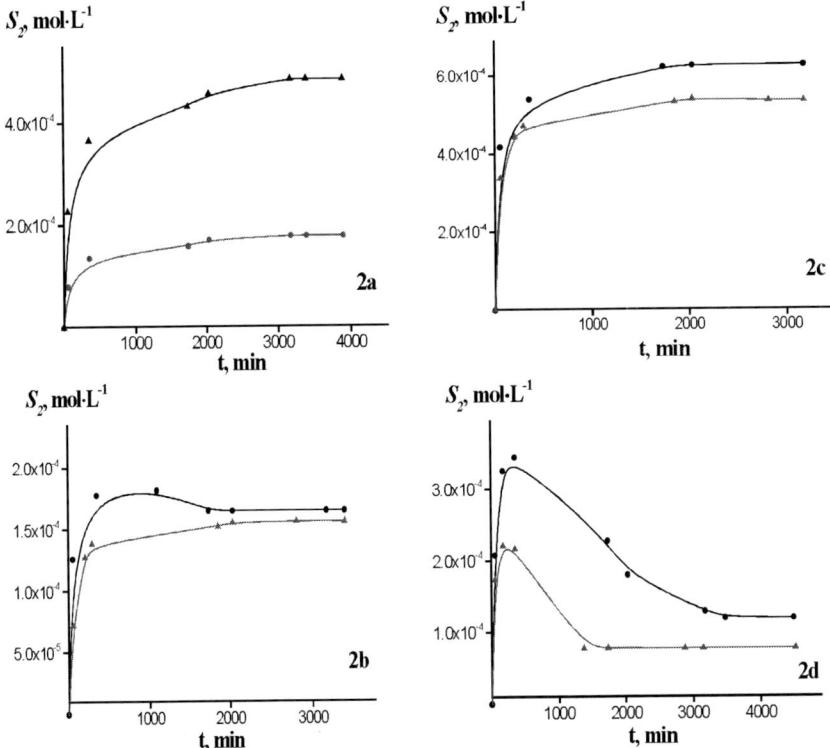

Figure 8. Kinetic solubility curves compounds **2a-d** in buffers pH 2.0 (●) and pH 7.4 (▲)

Based on the kinetic solubility data, it was determined that it took more than 60 hours for the studied systems to reach equilibrium.

Thermodynamic Solubility

The solubility of the synthesized derivatives in the stable equilibrium conditions or thermodynamic solubility was determined within the temperature range of (293.15-313.15) K by the shake-flask method. The results of the solubility measurements of compounds 2 (a - d) in the buffers (pH 2.0 and 7.4) and 1-octanol are shown in Table 10. Figure 9 presents the temperature dependences of solubility of the derivatives in the solvents used. The substance solubility was found to increase with temperature growth. The substance solubility values in the aqueous solvents did not exceed $9 \cdot 10^{-4}$ mol·L^{-1}. The main cause of the low solubility of the studied derivatives in water was the high hydrophobicity of the aromatic groups. When substituents were introduced, the solubility decreased in the following order: methoxy-(**2b**) > methyl-(**2a**) > chloro-(**2c**) > fluoro-(**2d**). It should be said that the halogen atoms produce a rather strong negative inductive effect, which may be the cause of the solubility decrease. The solubility in buffer pH 2.0 is higher than that in buffer pH 7.4 for all the compounds except chlorine derivative **2c** at 308.15 and 313.15 K.

Table 10. Temperature dependences of solubility S_2 (mol·L^{-1}) of the compounds studied in the buffer solutions (pH 2.0 and 7.4) and 1-octanol

T/K	2a			2b		
	buffer pH 2.0	buffer pH 7.4	1-octanol	buffer pH 2.0	buffer pH 7.4	1-octanol
	$S_2 \cdot 10^4$	$S_2 \cdot 10^4$	$S_2 \cdot 10^3$	$S_2 \cdot 10^4$	$S_2 \cdot 10^4$	$S_2 \cdot 10^2$
293.15	4.52	1.58	1.89	5.67	4.96	1.15
298.15	4.89	1.78	2.32	6.28	5.35	1.33
303.15	5.30	2.01	2.79	7.01	5.74	1.58
308.15	5.77	2.26	3.32	7.76	6.11	1.85
313.15	6.23	2.52	3.97	8.56	6.56	2.15
T/K	2c			2d		
	buffer pH 2.0	buffer pH 7.4	1-octanol	buffer pH 2.0	buffer pH 7.4	1-octanol
	$S_2 \cdot 10^4$	$S_2 \cdot 10^4$	$S_2 \cdot 10^3$	$S_2 \cdot 10^4$	$S_2 \cdot 10^5$	$S_2 \cdot 10^3$
293.15	1.57	1.45	3.47	1.12	7.18	2.12
298.15	1.65	1.56	3.89	1.19	7.62	2.41
303.15	1.71	1.67	4.39	1.27	8.00	2.76
308.15	1.78	1.79	4.83	1.35	8.44	3.10
313.15	1.85	1.92	5.36	1.44	8.94	3.51

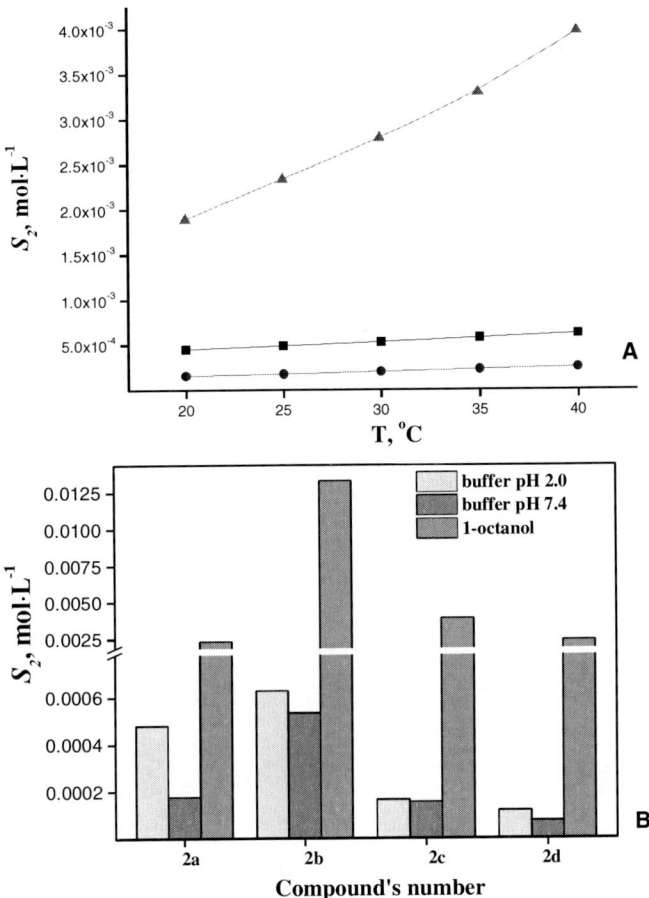

Figure 9. Solubility data of the compounds of group 2 in the selected solvents: A – Temperature dependence of equilibrium solubility for compound **2a** in the solvents studied: ■ – buffer pH 2.0, ● – buffer pH 7.4, ▲ –1-octanol; B – solubility histogram at 298.15 K.

The structure of the studied substances includes twelve proton acceptor atoms (nitrogen, oxygen, fluorine, and sulfur) and only the hydroxyl and amino-groups are proton donors. That is why the studied derivatives have basic properties, which is confirmed by the calculations made in the pDISOL-X program (Avdeef, 2014). The pKa values of the compounds are equal to 2.3, which means that buffer pH 2.0 contains ionized molecular forms and, as a result, the solubility in this buffer becomes higher. Drug compounds with high solubility in 1-octanol are known to dissolve well in nonpolar media of the body, such as lipids and the nervous system (Domańska et al. 2009). The results of the solubility measurements of the synthesized substances in 1-octanol are shown in Table 10. The solubility values of the substances in 1-octanol within the studied temperature interval range from $1.8 \cdot 10^{-3}$ to $2.2 \cdot 10^{-2}$ mol·L^{-1}. The higher solubility of the studied heterocyclic compounds in 1-octanol in comparison with that in the aqueous solutions is probably caused by the formation of hydrogen bonds with the amphiphilic alcohol molecules. The introduction of a methyl group and halogen atoms into the para-position of the phenyl ring does not cause considerable changes in the compound solubility. At the same time, the solubility of derivative **2b** in 1-octanol with a methoxy-substituent is almost an order of magnitude higher than in the other compounds. Thus, it is established that the solubility of the synthesized derivative with a methoxy-substituent in the ortho-position of the phenyl ring in the aqueous medium and 1-octanol is higher, which is probably the result of its ability to prevent the formation of solvent associates connected by a hydrogen bond rather than of the direct electronic effect.

Dissolution Thermodynamics

Thermodynamic functions provide important information about the dissolution processes of the compounds under study. The apparent standard dissolution enthalpy (ΔH_{sol}^{0}) in the selected solvent can be determined from the van't Hoff analysis (Eqs. (7)). The parameters of the equation $\ln x = A - B(1/T - 1/T_m)$ are given in Table 11. The apparent standard Gibbs energy (ΔG_{sol}^{0}) and entropy (ΔS_{sol}^{0}) were obtained from Eqs. (9) and (10), respectively (Kim et al. 2018): We calculated the apparent thermodynamic functions of dissolution of the studied compounds in the buffer solutions and 1-octanol by equations (7-10) based on the solubility temperature dependences (Table 12).

The obtained values of the Gibbs energy, enthalpy, and entropy can be used to compare the substance dissolution in aqueous and organic solvents. The Gibbs energy values in all the studied systems are positive, which indicates that the dissolution process is not favorable. And the ΔG_{sol}^{0} value in this case decreases depending on the solvent used in the following order: buffer pH 7.4 > buffer pH 2.0 > 1-octanol, which agrees with the solubility increase in the substances. The enthalpy values of the compounds are positive, which means that the dissolution process is endothermic.

Table 11. Coefficients of the correlation equation $\ln x_2 = A - B/(1/T - 1/T_{hm})$

Compound	A	B	R
buffer pH 2.0			
2a	-11.5 ± 0.02	1513 ± 22	0.9997
2b	-11.3 ± 0.01	1929 ± 17	0.9999
2c	-12.7 ± 0.01	773 ± 17	0.9993
2d	-13.0 ± 0.01	1181 ± 16	0.9997
buffer pH 7.4			
2a	-12.5 ± 0.01	2187 ± 17	0.9999
2b	-11.5 ± 0.01	1301 ± 17	0.9997
2c	-12.7 ± 0.01	1321 ± 15	0.9998
2d	-13.4 ± 0.01	1023 ± 21	0.9994
1-octanol			
2a	-7.7 ± 0.002	3457 ± 29	0.9995
2b	-6.0 ± 0.003	2993 ± 47	0.9996
2c	-7.3 ± 0.002	2069 ± 28	0.9997
2d	-7.7 ± 0.002	2384 ± 24	0.9998

Table 12. Apparent thermodynamic solubility functions of the compounds of group 2 in buffer pH 2.0, buffer pH 7.4 and 1-octanol at 298.15 K

Compound	ΔG^0_{sol} kJ·mol^{-1}	ΔH^0_{sol} kJ·mol^{-1}	$T\Delta S^0_{sol}$ kJ·mol^{-1}	ΔS^0_{sol} J·mol^{-1}·K^{-1}
	buffer pH 2.0			
2a	29.0± 0.6	12.6 ± 0.2	-16.4	-54.1± 0.6
2b	28.4± 0.5	16.0 ± 0.1	-12.4	-40.9± 0.5
2c	32.0 ± 0.6	6.4 ± 0.1	-25.6	-84.5± 0.4
2d	32.7 ± 0.7	9.8 ± 0.1	-22.9	-75.6± 0.4
	buffer pH 7.4			
2a	31.5 ± 0.6	18.2 ± 0.1	-13.3	-43.9 ±0.7
2b	29.0 ± 0.6	10.8 ± 0.1	-18.2	-60.1 ±0.5
2c	32.0 ± 0.6	11.0 ± 0.1	-21.0	-69.3 ±0.4
2d	33.7 ± 0.6	8.5 ± 0.2	-25.2	-83.2 ±0.5
	1-octanol			
2a	19.4 ± 0.4	28.7 ± 0.5	9.3	30.7± 1.9
2b	15.1 ± 0.5	24.9 ± 0.1	9.8	32.4 ± 1.2
2c	18.4± 0.3	17.2 ± 0.2	-1.2	-4.0 ± 0.7
2d	19.4 ± 0.4	19.8 ± 0.2	0.4	1.3 ± 0.1

The entropy of dissolution is negative in the buffer solutions and positive and slightly negative in 1-octanol. The enthalpy and entropy contributions to the Gibbs energy were evaluated by the diagram approach (Figure 10).

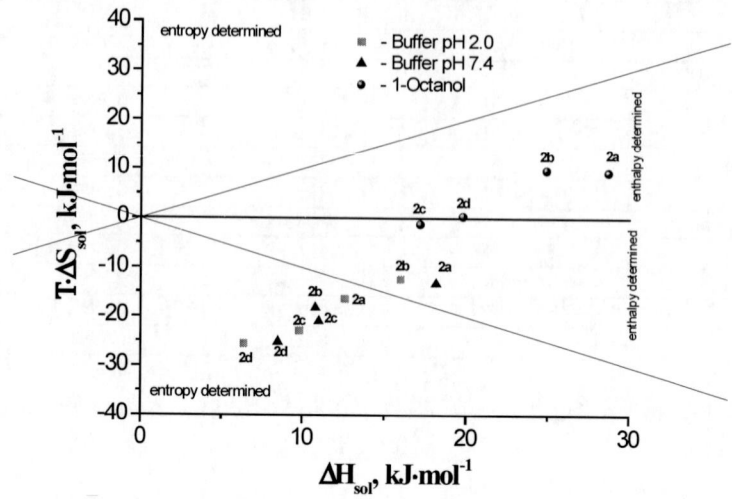

Figure 10. Relationship between the thermodynamic functions of solubility processes in buffers (pH 2.0 and 7.4), 1-octanol for the derivatives studied.

As shown from figure 10 data the ΔH^0_{sol} and $T\Delta S^0_{sol}$ values compensate for each other (R=0.957). The dissolution process in the aqueous medium is entropy-determined (in absolute value) ($\Delta H^0_{sol} < T\Delta S^0_{sol}$), whereas in 1-octanol it is controlled by enthalpy term ($\Delta H^0_{sol} > T\Delta S^0_{sol}$). The only exception is the following systems: compound **2c** - buffer pH 2.0 and compound **2a** - buffer pH 7.4. The entropy increases the Gibbs energy in the buffer solutions and lowers the solubility. On the contrary, the entropy contribution in 1-octanol reduces the Gibbs energy and increases the solubility of the studied compounds.

Hansen Solubility Parameters

Hansen solubility parameters (HSP) were developed to predict materials compatibilities in various fields of study, including in pharmaceutical drug development (Martínez, Peña and Bustamante, 2011, Mohammad, Alhalaweh and Velaga, 2011, Breitkreutz, 1998). The method is based on the idea that *"like dissolves like."* This is the case when the solvent and the solute have similar HSP. The three HSP quantitatively represent nonpolar (atomic) bonding, permanent dipole (molecular) bonding, and hydrogen (molecular) bonding. The basic equation governing the assignment of Hansen parameters is the sum of the individual energies that make it up:

$$\delta_t^2 = \delta_d^2 + \delta_p^2 + \delta_h^2 \qquad (19)$$

where δ_d, δ_p and δ_h are the HSPs corresponding to dispersion, polar and hydrogen bonding forces, respectively. The partial solubility parameters are calculated by the following equations:

$$\delta_d = \Sigma F_{di}/\Sigma V_i \qquad (20)$$

$$\delta_p = (\Sigma F_{pi}^2)^{0.5}/\Sigma V_i; \qquad (21)$$

$$\delta_h = (\Sigma F_{hi}/\Sigma V_i)^{0.5} \qquad (22)$$

where V_i denotes the contribution of the molar group i to the molar volume of the molecule and F_{di}, F_{pi} and E_{hi} are defined as contributions of the dispersion forces, polarity and hydrogen bond structural group i to the cohesion energy.

The difference between the calculated parameters of solute (1) and solvent (2) was determined according to Eq. (15) as the Euclidean distance or radius interaction:

$$\Delta\delta = ((\delta_{d1} - \delta_{d2})^2 + (\delta_{p1} - \delta_{p2})^2 + (\delta_{h1} - \delta_{h2})^2)^{0.5} \qquad (23)$$

Ideally, miscibility is taken into account for a solution containing a constant amount of a dissolved substance. Besides, Bagley established a new conceptual thermodynamic connection between δ_d and δ_p and introduced a combined solubility parameter δ_V:

$$\delta_v = (\delta_d^2 + \delta_p^2)^{1/2} \qquad (24)$$

The partial solubility parameters for the compounds studied and selected solvents were estimated using Van Krevelen–Hoftyzer's and Fedors's combined group contribution methods (Van Krevelen and Hoftyzer, 1976; Fedors, 1974). The group contribution parameters and associated molar volumes of the derivatives are summarized in Table 13.

Table 13. Group contribution parameters and associated molar volumes of compounds 2a-d

Individual functional group	Frequency	F_{di}, $(J/cm^3)^{0.5} \cdot mol^{-1}$	F_{pi}, $(J/cm^3)^{0.5} \cdot mol^{-1}$	E_{hi}, J/mol	V_i, cm^3/mol
2a					
-F	2	102.0	493.9	6544.3	18.0
-CH$_3$	1	336.6	0	0	33.5
-CH$_2$-	2	234.6	0	0	16.1
>C<	2	-214.2	0	0	-19.2
-OH	1	76.5	1225.0	6060.0	10.0
-CO-	2	105	600.0	9500.0	10.8
-NH-	1	122.4	700.7	1500.0	4.5
-S-	2	815.9	196.0	297.5	12.0
=C<	3	56.7	2.00	0	5.5
-N=	3	380	100	250	5.0
>N-	3	30	150.0	750.0	-9
Ring closure 5 or more atoms	3	142.8	0	0	16
Phenylene (o, m, p)	2	1173.0	63.7	40.4	52.4
Double bond	1	15.0	14.3	83.5	2.2
2b					
-F	2	102.0	493.9	6544.3	18.0
-CH$_3$	1	336.6	0	0	33.5
-O-	1	76.5	1225.0	101.0	3.8
-CH$_2$-	2	234.6	0	0	16.1
>C<	2	-214.2	0	0	-19.2
-OH	1	76.5	1225.0	6060.0	10.0
-CO-	2	105	600.0	9500.0	10.8
-NH-	1	122.4	700.7	1500.0	4.5
-S-	2	815.9	196.0	297.5	12.0
=C<	3	56.7	2.00	0	5.5
-N=	3	380	100	250	5.0
>N-	3	30	150.0	750.0	-9
Ring closure 5 or more atoms	3	142.8	0	0	16
Phenylene (o, m, p)	2	1173.0	63.7	40.4	52.4
Double bond	1	15.0	14.3	83.5	2.2

Individual functional group	Frequency	F_{di}, $(J/cm^3)^{0.5} \cdot mol^{-1}$	F_{pi}, $(J/cm^3)^{0.5} \cdot mol^{-1}$	E_{hi}, J/mol	V_i, cm³/mol
2c					
F	2	102.0	493.9	6544.3	18.0
Cl	1	397.8	1477.2	4706.0	26.0
-CH$_2$-	2	234.6	0	0	16.1
>C<	2	-214.2	0	0	-19.2
-OH	1	76.5	1225.0	6060.0	10.0
-CO-	2	105	600.0	9500.0	10.8
-NH-	1	122.4	700.7	1500.0	4.5
-S-	2	815.9	196.0	297.5	12.0
=C<	3	56.7	2.00	0	5.5
-N=	3	380	100	250	5.0
>N-	3	30	150.0	750.0	-9
Ring closure 5 or more atoms	3	142.8	0	0	16
Phenylene (o, m, p)	2	1173.0	63.7	40.4	52.4
Double bond	1	15.0	14.3	83.5	2.2
2d					
-F	3	102.0	493.9	6544.3	18.0
-CH$_2$-	2	234.6	0	0	16.1
>C<	2	-214.2	0	0	-19.2
-OH	1	76.5	1225.0	6060.0	10.0
-CO-	2	105	600.0	9500.0	10.8
-NH-	1	122.4	700.7	1500.0	4.5
-S-	2	815.9	196.0	297.5	12.0
=C<	3	56.7	2.00	0	5.5
-N=	3	380	100	250	5.0
>N-	3	30	150.0	750.0	-9
Ring closure 5 or more atoms	3	142.8	0	0	16
Phenylene (o, m, p)	2	1173.0	63.7	40.4	52.4
Double bond	1	15.0	14.3	83.5	2.2

The HSP parameters of the compounds and solvents, as well as the calculated molar volumes at 298.15 K are presented in Table 14.

According to the "similarity and intermiscibility" theory, compounds with similar solubility values are characterized by stronger intermolecular interactions, which makes the dissolution process more effective (Greenhalgh et al., 1999). An analysis of the data in Table 14 shows that the $\Delta\delta$ values of all the studied substances in 1-octanol are much lower than in the buffer solutions. The calculated values of the $\Delta\delta$ parameter agree with the experimentally determined solubility of the compounds demonstrating better miscibility with the alcohol than with the water. Of certain interest is also the fact that the δ_d parameters in the buffer solutions and 1-octanol are quite close and, therefore, the contributions of the dispersion forces to the dissolution of the compounds in the aqueous and organic solvents do not differ radically either. On the contrary, the higher values of the δ_p and δ_h parameters in the buffer solutions in comparison with those in 1-octanol lead to a considerable difference in the $\Delta\delta$ values. Based on the obtained results, we can conclude that the dipole-dipole interactions and hydrogen bonding between the solute and the solvent are the main forces driving the dissolution of the studied compounds.

Table 14. Molar volumes and Hansen solubility parameters for the compounds studied in the selected solvents

Compound	V, cm³/mol	δ_d MPa$^{0.5}$	δ_p MPa$^{0.5}$	δ_h MPa$^{0.5}$	δ_t MPa$^{0.5}$	$\Delta\delta_t$ MPa$^{0.5}$	$\Delta\delta$ MPa$^{0.5}$	δ_v MPa$^{0.5}$
2a	282.9	24.3	5.8	12.4	27.9	-	-	25.0
buffer solutions	18.0	15.5	16.0	42.3	47.8	19.9	32.8	4.0
1-octanol	157.7	17.0	3.3	11.9	21.0	6.9	7.7	16.7
2b	286.7	24.2	7.1	12.3	28.1	-	-	25.2
buffer solutions	18.0	15.5	16.0	42.3	47.8	19.7	32.5	4.0
1-octanol	157.7	17.0	3.3	11.9	21.0	7.1	8.1	16.7
2c	275.4	24.7	8.0	13.2	29.1	-	-	26.0
buffer solutions	18.0	15.5	16.0	42.3	47.8	18.7	31.5	4.0
1-octanol	157.7	17.0	3.3	11.9	21.0	8.1	9.1	16.7
2d	267.4	24.8	6.1	13.7	29.0	-	-	25.5
buffer solutions	18.0	15.5	16.0	42.3	47.8	18.8	31.7	4.0
1-octanol	157.7	17.0	3.3	11.9	21.0	8.0	8.5	16.7

The previous investigations showed that the systems with $\Delta\delta_t > 10$ MPa$^{0.5}$ had low miscibility, whereas those with $\Delta\delta_t$ in the range of 7-10 MPa$^{0.5}$ were partially miscible (Mohammad, Alhalaweh, and Velaga, 2011). The δ_t values in the buffer solutions were higher than 10 MPa$^{0.5}$, which might be the cause of the low solubility of the substances in the aqueous solvent. At the same time, the $\Delta\delta_t$ values in 1-octanol did not exceed 8.1 MPa$^{0.5}$, which agreed with the comparatively higher solubility of the compounds in the alcohol.

Modeling of Solubility Data

In order to quantitatively describe the relationship between the experimental solubility and the temperature was made modeling by the modified Apelblat and van't Hoff equations. The mole fraction solubilities of the studied compounds calculated by Eqs. (1-3) are reported in Table 15. The model parameters with the values of deviations calculated by Eqs. (4-6) are presented in Table 16.

Table 15. Experimental (x_2^{exp}) and correlated (x_2^{cal}) mole fractions of solubility for compounds 2a-d in the selected solvents at different temperatures

T/K	x_2^{exp}	Modified Apelblat equation		van't Hoff equation	
		x_2^{cal}	RD[a]	x_2^{cal}	RD[a]
2a					
buffer pH 2.0					
293.15	8.14·10⁻⁶	8.14·10⁻⁶	0.0003	8.11·10⁻⁶	0.0044
298.15	8.83·10⁻⁶	8.83·10⁻⁶	-0.0004	8.85·10⁻⁶	-00020
303.15	9.58·10⁻⁶	9.59·10⁻⁶	-0.0015	9.62·10⁻⁶	-0.0046
308.15	10.44·10⁻⁶	10.42·10⁻⁶	0.0022	10.43·10⁻⁶	0.0008
313.15	11.31·10⁻⁶	11.32·10⁻⁶	-0.0010	11.28·10⁻⁶	0,0024
buffer pH 7.4					
293.15	2.83·10⁻⁶	2.83·10⁻⁶	0.0012	2.82·10⁻⁶	0.0022
298.15	3.19·10⁻⁶	3.20·10⁻⁶	-0.0024	3.20·10⁻⁶	-0.0027
303.15	3.61·10⁻⁶	3.61·10⁻⁶	-0.0009	3.61·10⁻⁶	-0.0016
308.15	4.07·10⁻⁶	4.06·10⁻⁶	0.0036	4.06·10⁻⁶	0.0034
313.15	4.54·10⁻⁶	4.55·10⁻⁶	-0.0018	4.55·10⁻⁶	-0.0008
1-octanol					
293.15	2.99·10⁻⁴	2.99·10⁻⁴	-0.0025	3.00·10⁻⁴	-0.0045
298.15	3.68·10⁻⁴	3.66·10⁻⁴	0.0049	3.66·10⁻⁴	0.0061
303.15	4.44·10⁻⁴	4.44·10⁻⁴	-0.0001	4.43·10⁻⁴	0.0020
308.15	5.31·10⁻⁴	5.33·10⁻⁴	-0.0052	5.33·10⁻⁴	-0.0042
313.15	6.38·10⁻⁴	6.36·10⁻⁴	0.0025	6.38·10⁻⁴	0.0005
2b					
buffer pH 2.0					
293.15	1.02·10⁻⁵	1.02·10⁻⁵	0.0012	1.02·10⁻⁵	0.0027
298.15	1.13·10⁻⁵	1.14·10⁻⁵	-0.0033	1.14·10⁻⁵	-0.0043
303.15	1.27·10⁻⁵	1.26·10⁻⁵	0.0019	1.27·10⁻⁵	0.0001
308.15	1.40·10⁻⁵	1.40·10⁻⁵	0.0007	1.40·10⁻⁵	-0.0001
313.15	1.55·10⁻⁵	1.55·10⁻⁵	-0.0007	1.55·10⁻⁵	0.0007
buffer pH 7.4					
293.15	8.89·10⁻⁶	8.89·10⁻⁶	-0.0008	8.90·10⁻⁶	-0.0010
298.15	9.60·10⁻⁶	9.58·10⁻⁶	0.0018	9.58·10⁻⁶	0.0015
303.15	10.32·10⁻⁶	10.30·10⁻⁶	0.0018	10.30·10⁻⁶	0.0015
308.15	11.00·10⁻⁶	11.04·10⁻⁶	-0.0038	11.05·10⁻⁶	-0.0040
313.15	11.84·10⁻⁶	11.82·10⁻⁶	0.0019	11.82·10⁻⁶	0.0018
1-octanol					
293.15	1.82·10⁻³	1.82·10⁻³	0.0035	1.81·10⁻³	0.0080
298.15	2.12·10⁻³	2.14·10⁻³	-0.0095	2.14·10⁻³	-0.0116
303.15	2.53·10⁻³	2.52·10⁻³	0.0049	2.53·10⁻³	0.0008
308.15	2.97·10⁻³	2.97·10⁻³	0.0024	2.97·10⁻³	0.0007
313.15	3.48·10⁻³	3.49·10⁻³	-0.0025	3.47·10⁻³	0.0022

Table 15. (Continued)

T/K	x_2^{exp}	Modified Apelblat equation		van't Hoff equation	
		x_2^{cal}	RD[a]	x_2^{cal}	RD[a]
2c					
buffer pH 2.0					
293.15	2.83·10⁻⁶	2.83·10⁻⁶	-0.0010	2.84·10⁻⁶	-0.0030
298.15	2.97·10⁻⁶	2.97·10⁻⁶	0.0021	2.97·10⁻⁶	0.0036
303.15	3.10·10⁻⁶	3.10·10⁻⁶	-0.0010	3.09·10⁻⁶	0.0016
308.15	3.23·10⁻⁶	3.23·10⁻⁶	-0.0011	3.22·10⁻⁶	0.0002
313.15	3.35·10⁻⁶	3.35·10⁻⁶	0.0006	3.36·10⁻⁶	-0.0015
buffer pH 7.4					
293.15	2.59·10⁻⁶	2.59·10⁻⁶	-0.0005	2.59·10⁻⁶	-0.0001
298.15	2.80·10⁻⁶	2.79·10⁻⁶	0.0016	2.80·10⁻⁶	0.0009
303.15	3.01·10⁻⁶	3.01·10⁻⁶	0.0017	3.01·10⁻⁶	0.0007
308.15	3.22·10⁻⁶	3.23·10⁻⁶	-0.003	3.23·10⁻⁶	-0.0036
313.15	3.47·10⁻⁶	3.46·10⁻⁶	0.0017	3.46·10⁻⁶	0.0023
1-octanol					
293.15	5.48·10⁻⁴	5.48·10⁻⁴	0.0005	5.49·10⁻⁴	-0.0017
298.15	6.17·10⁻⁴	6.19·10⁻⁴	-0.0034	6.18·10⁻⁴	-0.0017
303.15	6.99·10⁻⁴	6.96·10⁻⁴	0.0047	6.94·10⁻⁴	0.0075
308.15	7.73·10⁻⁴	7.76·10⁻⁴	-0,0038	7.75·10⁻⁴	-0.0023
313.15	8.61·10⁻⁴	8.61·10⁻⁴	0.0006	8.63·10⁻⁴	-0.0016
2d					
buffer pH 2.0					
293.15	2.01·10⁻⁶	2.01·10⁻⁶	-0.0007	2.01·10⁻⁶	0.0014
298.15	2.15·10⁻⁶	2.15·10⁻⁶	0.0016	2.15·10⁻⁶	0.0006
303.15	2.29·10⁻⁶	2.29·10⁻⁶	0.0002	2.29·10⁻⁶	-0.0018
308.15	2.44·10⁻⁶	2.44·10⁻⁶	-0.0018	2.45·10⁻⁶	-0.0027
313.15	2.61·10⁻⁶	2.61·10⁻⁶	0.0010	2.60·10⁻⁶	0.0032
buffer pH 7.4					
293.15	1.29·10⁻⁶	1.29·10⁻⁶	-0.0013	1.29·10⁻⁶	0.0003
298.15	1.37·10⁻⁶	1.36·10⁻⁶	0.0042	1.36·10⁻⁶	0.0029
303.15	1.44·10⁻⁶	1.44·10⁻⁶	-0.0013	1.44·10⁻⁶	-0.0035
308.15	1.52·10⁻⁶	1.52·10⁻⁶	-0.0015	1.52·10⁻⁶	-0.0027
313.15	1.61·10⁻⁶	1.61·10⁻⁶	0.0016	1.61·10⁻⁶	0.0033
1-octanol					
293.15	3.35·10⁻⁴	3.34·10⁻⁴	0.0012	3.34·10⁻⁴	0.0012
298.15	3.82·10⁻⁴	3.83·10⁻⁴	-0.0033	3.83··10⁻⁴	-0.0037
303.15	4.39·10⁻⁴	4.37·10⁻⁴	0.0055	4.37·10⁻⁴	0.0050
308.15	4.95·10⁻⁴	4.97·10⁻⁴	-0.0031	4.97·10⁻⁴	-0.0034
313.15	5.63·10⁻⁴	5.62··10⁻⁴	0.0011	5.62··10⁻⁴	0.0011

[a] RD is the relative deviation: $RD = (x_{exp} - x_{cal})/x_{exp}$

Standard uncertainities: $u(T) = 0.15$ K and $u(p) = 3$ kPa.

Relative standard uncertainties for solubility: $u_r(x) = 0.045$ for buffer solutions and $u_r(x) = 0.04$ for hexane and 1-octanol.

Table 16. Parameters of the modified Apelblat and van't Hoff equations for the compounds studied in the selected solvents

Solvents	A	B	C	RMSD	RAD
2a					
Modified Apelblat equation					
buffer pH 2.0	-85.98	2068.9	11.83	$1.73 \cdot 10^{-8}$	0.0011
buffer pH 7.4	-25.38	-1283.8	2.99	$8.50 \cdot 10^{-9}$	0.0020
1-octanol	55.14	-5782.0	-7.66	$1.68 \cdot 10^{-6}$	0.0030
van't Hoff equation					
buffer pH 2.0	-6.55	-1514.5	-	$2.80 \cdot 10^{-8}$	0.0027
buffer pH 7.4	-5.31	-2189.4	-	$8.33 \cdot 10^{-9}$	0.0021
1-octanol	3.69	-3461.1	-	$1.59 \cdot 10^{-6}$	0.0035
2b					
Modified Apelblat equation					
buffer pH 2.0	-44.76	-132.7	5.94	$2.17 \cdot 10^{-8}$	0.0015
buffer pH 7.4	-9.29	-1208.1	0.31	$2.41 \cdot 10^{-9}$	0.0020
1-octanol	-103.08	1830.7	15.93	$1.21 \cdot 10^{-5}$	0.0046
van't Hoff equation					
buffer pH 2.0	-4.91	-1930.9	-	$2.56 \cdot 10^{-8}$	0.0020
buffer pH 7.4	-7.18	-1303.1	-	$2.42 \cdot 10^{-8}$	0.0020
1-octanol	3.90	-2996.2	-	$1.33 \cdot 10^{-5}$	0.0047
2c					
Modified Apelblat equation					
buffer pH 2.0	47.22	-3361.9	-8.54	$3.76 \cdot 10^{-9}$	0.0012
buffer pH 7.4	-27.75	-447.0	2.89	$5.95 \cdot 10^{-9}$	0.0017
1-octanol	62.07	-4891.4	-9.31	$2.19 \cdot 10^{-6}$	0.0026
van't Hoff equation					
buffer pH 2.0	-10.13	-774.2	-	$6.87 \cdot 10^{-9}$	0.0019
buffer pH 7.4	-8.35	-1322.4	-	$6.44 \cdot 10^{-9}$	0.0015
1-octanol	-0.44	-2071.0	-	$2.63 \cdot 10^{-6}$	0.0029
2d					
Modified Apelblat equation					
buffer pH 2.0	-60.79	1150.5	7.70	$2.82 \cdot 10^{-9}$	0.0011
buffer pH 7.4	-58.44	1158.1	7.21	$3.19 \cdot 10^{-9}$	0.0020
1-octanol	-5.90	-2114.8	0.90	$1.43 \cdot 10^{-6}$	0.0028
van't Hoff equation					
buffer pH 2.0	-9.08	-1182.5	-	$5.24 \cdot 10^{-9}$	0.0019
buffer pH 7.4	-10.07	-1024.4	-	$4.14 \cdot 10^{-9}$	0.0025
1-octanol	0.14	-2387.3	-	$1.43 \cdot 10^{-6}$	0.0029

As the data show, the modeling results obtained by the modified Apelblat and van't Hoff equations show good consistency with the experimental solubility with RAD values less than 1% and these models are all suitable for correlating the solubility of the compounds studied in the selected solvents.

Obviously, the modified Apelblat equation has the smallest values of deviations for the investigated systems.

Partition in the 1-Octanol/Buffer pH 7.4 System

The 1-octanol/buffer pH 7.4 partition coefficients of the studied compounds were determined at different temperatures by the shake-flask method. The $P_{O/B}$ values of the derivatives were calculated from the equilibrium concentrations expressed in molarity in the 1-octanol and aqueous phases. The 1-octanol/buffer pH 7.4 partition coefficients of the studied compounds within the temperature range of (293.15 - 313.15) K are given in Table 17 and shown in Figure 11. The data obtained indicate that the log$P_{O/B}$ values of the compounds ranged from 0.80 to 1.27 at 298.15 K and increased as the temperature rises. The general rule for achieving optimal gastrointestinal absorption by passive diffusion permeability after oral dosing is to have a moderate log$P_{O/B}$ range (0–3). This range is characterized by a good balance between the permeability and solubility (Kerns and Di, 2008). The log$P_{O/B}$ values reflect the displacement of the equilibrium in the studied systems from the water phase to the octanol one, and this displacement increases with temperature growth. The partition coefficients of the substances decrease in the following order when the substituents are changed: -CH$_3$(**2a**) > -Cl(**2c**) > -OCH$_3$(**2b**) $\geq$ –F(**2d**).

The established trend of decreasing log$P_{O/B}$ values corresponds to the change in the chemical nature of the introduced substituents from the electron-donating methyl to the electron-acceptor chlorine and fluorine atoms. The highest partition coefficients of compound **2a** are explained by the increase in the molecule hydrophobicity after the methyl group introduction into the phenyl ring (Harrold and Zavod, 2013). The chlorine derivative has higher lipophilicity than the fluorinated compound, which agrees with the data obtained earlier (Hallgas et al., 2007).

The compounds with a methoxy-substituent and a fluorine atom have practically the same partition coefficient values. The introduction of a methoxy-group may not only lead to changes in the molecule conformation and increase in its polarity, but also change the water structure in the compound hydration shell, which is a factor that reduces the molecule hydrophobicity and lipophilicity. It should be said that the partition

Table 17. Experimental concentrations and partition coefficients of the studied compounds in 1-octanol/buffer pH 7.4 system at different temperatures

T/K	2a						2b					
	$s_O \cdot 10^4$	$s_B \cdot 10^6$	$\log P_{O/B}$	$x_O \cdot 10^5$	$x_B \cdot 10^7$	$\log P'_{O/B}$	$s_O \cdot 10^4$	$s_B \cdot 10^5$	$\log P_{O/B}$	$x_O \cdot 10^5$	$x_B \cdot 10^7$	$\log P'_{O/B}$
293.15	1.59	9.54	1.22	2.51	1.71	2.17	1.24	2.08	0.78	1.96	3.72	1.72
298.15	1.60	8.44	1.27	2.54	1.51	2.22	1.25	1.97	0.80	1.99	3.53	1.75
303.15	1.61	7.63	1.32	2.56	1.37	2.27	1.26	1.88	0.83	2.01	3.37	1.77
308.15	1.62	6.81	1.38	2.59	1.22	2.32	1.27	1.78	0.86	2.03	3.20	1.80
313.15	1.63	6.17	1.42	2.61	1.11	2.37	1.28	1.69	0.88	2.06	3.05	1.83
A^a	12.3 ± 0.09						7.82 ± 0.04					
B^a	2135 ± 28						1132 ± 13					
R^b	0.9997						0.9998					

T/K	2c						2d					
	$s_O \cdot 10^4$	$s_B \cdot 10^6$	$\log P_{O/B}$	$x_O \cdot 10^5$	$x_B \cdot 10^7$	$\log P'_{O/B}$	$s_O \cdot 10^4$	$s_B \cdot 10^5$	$\log P_{O/B}$	$x_O \cdot 10^5$	$x_B \cdot 10^7$	$\log P'_{O/B}$
293.15	3.52	2.64	1.13	5.55	4.72	2.07	1.77	3.00	0.77	2.80	5.37	1.72
298.15	3.54	2.48	1.16	5.60	4.44	2.10	1.79	2.85	0.79	2.83	5.10	1.74
303.15	3.56	2.31	1.19	5.63	4.15	2.13	1.80	2.71	0.82	2.85	4.86	1.77
308.15	3.57	2.20	1.21	5.69	3.96	2.16	1.81	2.61	0.84	2.89	4.69	1.79
313.15	3.58	2.07	1.24	5.74	3.73	2.19	1.83	2.50	0.86	2.92	4.50	1.81
A^a	8.92 ± 0.04						7.41 ± 0.06					
B^a	1217 ± 13						1013 ± 18					
R^b	0.9998						0.9995					

[a] parameters of the correlation equation: $\log P'_{O/B} = A - B/T$;

[b] R is the pair correlation coefficient;

[c] σ is the standard deviation.

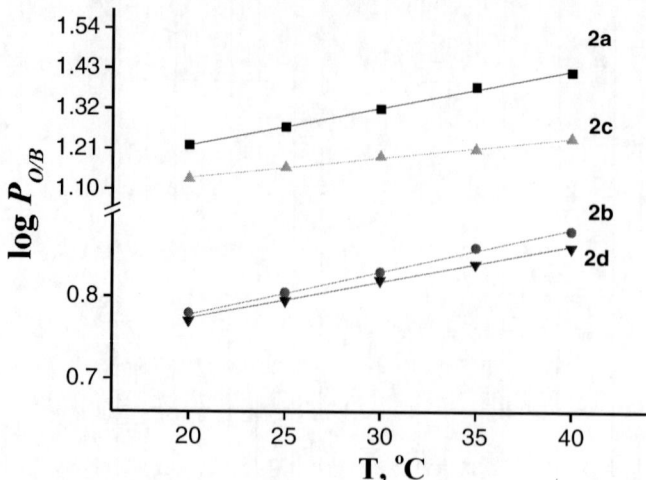

Figure 11. Partition coefficients of the compounds of group 2 in the 1-octanol/buffer pH 7.4 system at different temperatures.

coefficients obtained experimentally cannot be calculated based on the solubility data and, vice versa, the solubility in one of the phases cannot be estimated based on the known partition coefficient (Józan, Takácsné-Novak, and Szász, 1996).

Transfer Thermodynamics

We calculated the thermodynamic functions of transfer (Table 18) based on the temperature dependence of the partition coefficients ($\log P^*_{O/B}$) of the studied compounds in the 1-octanol/buffer pH 7.4 system by equations (13-15).

The quantities $\Delta_{tr}H^o$ and $\Delta_{tr}S^o$ represent, respectively, the change in enthalpy and entropy when one mole of a solute is transferred from the aqueous phase to the 1-octanol phase (Dearden and Bresnen, 2005). The transfer Gibbs energies are negative in all the studied systems, which means that the transfer of the substances from the aqueous phase to the organic one is favourable. The positive enthalpy and entropy values allow us to conclude that the transfer is endothermic and entropy-driven. The substance transfer from the buffer solution to the 1-octanol medium is probably accompanied by energy absorption due to the molecule resolvation. Besides, the fact that the $T\Delta S^0_{tr} > \Delta H^0_{tr}$ inequality is not satisfied means that the entropy component makes the main contribution to the Gibbs energy of the transfer.

Table 18. Thermodynamic functions of transfer for the compounds studied at 298 K

Compound	ΔG_{tr}^o kJ·mol^{-1}	ΔH_{tr}^0 kJ·mol^{-1}	ΔS_{tr}^0 J·mol^{-1}·K^{-1}	$T\Delta S_{tr}^0$ kJ·mol^{-1}
2a	-12.8 ± 0.2	17.7 ± 0.2	102.1 ± 0.7	30.5
2b	-10.0 ± 0.2	9.4 ± 0.2	65.0 ± 0.3	19.4
2c	-12.0 ± 0.2	10.1 ± 0.1	74.2 ± 0.3	22.1
2d	-9.9 ± 0.2	8.4 ± 0.2	61.6 ± 0.5	18.3

Conclusion

To sum it up, a novel series of 10 derivatives of thiazolo[4,5-d]pyrimidines with (1H-1,2,4)- triazole were designed and synthesized using the molecular hybridization approach. The obtained substances had substituents of different chemical nature (methyl-, methoxy-, chloro-, and fluoro-) in their molecule phenyl ring. The chemical structures of the target compounds were characterized by 1H NMR spectroscopy. The PXRD technique and DSC method employed in the study showed that the solid forms of the substances were amorphous. An analysis was done of the effect produced by the linker structure and chemical nature of the substituents introduced into the phenyl ring on the antifungal activity of the compounds. Some of the derivatives containing an alkylpiperazinyl linker exhibited strong in vitro antifungal activity, comparable to fluconazole, against a broad spectrum of cultures: C. parapsilosis ATCC, 22019, C. albicans ATCC 24433, C. albicans 8R, C. albicans CBS 8837, C. albicans 604M, C. utilis 84, C. tropicalis 3019, C. glabrata 61L, Cryptococcus neoformans, and C. krusei 432M yeast cultures, A.niger 37a, filamentous fungi: M.canis B-200 and T.rubrum 2002.

The kinetic solubility of the synthesized substances was studied in aqueous solutions at physiological values of the gastro-intestinal tract medium acidity. The solubility of the synthesized derivatives in the stable equilibrium conditions was determined in the buffers (pH 2.0 and pH 7.4) and in 1-octanol within the temperature range of (293.15-313.15) K by the shake-flask method. The substance solubility in the aqueous solvents was found not to exceed 8.56·10-4 mol·L-1 and to decrease in the following order when the substituents were introduced: -CH3O, -CH3 >-Cl- > –F. Since the studied compounds are bases (pKa = 2.3), their solubility grew with an increase in the buffer solution acidity. The solubility values of the substances in 1-octanol within the studied temperature interval were higher than in the buffers and ranged from $1.8 \cdot 10{-3}$ to $5.8 \cdot 10{-2}$ mol·L-1. Hansen solubility parameters were applied to study the solubility behavior of the compounds in the selected solvents. A conclusion was made that the dipole-dipole interactions and hydrogen bonding were the main forces driving the dissolution of the studied substances. The solubility data were correlated by the van't Hoff and modified Apelblat equations. The smallest differences between the experimental and calculated solubility values were determined by the modified Apelblat equation. The apparent thermodynamic functions of dissolution of the studied compounds in the buffer solutions and 1-octanol were calculated based on solubility temperature dependences. The dissolution process in all the studied solute-solvent systems was endothermic. The enthalpy and entropy components of the dissolution Gibbs energy were analyzed by the diagram approach. A conclusion was made that in the aqueous medium, the dissolution process was controlled by the entropy term, and in 1-octanol - by the enthalpy one. The 1-octanol/buffer pH 7.4 partition coefficients of the compounds were determined within a temperature interval of (293.15 - 313.15) K by the shake-flask method. The results showed that the logPO/B values ranged from 0.47 to 1.19 at the temperature of 298.15 K and increased when the temperature rose. The obtained data allowed us to conclude that the synthesized derivatives had a good balance between the solubility and permeability. Based on the temperature dependences of the partition coefficients, we calculated the thermodynamic functions of transfer. The negative values of the transfer Gibbs energy and entropy mean that the substance transfer from the water phase to the organic one was a favorable, endothermic, and entropy-driven process.

References

Aggarwal, R. and Sumran, G. (2020). An insight on medicinal attributes of (1H-1,2,4) triazoles. *European Journal of Medicinal Chemistry*, 112652.

Atkins, P. and De Paula, J. (2006) *Physical chemistry for the life sciences,* W. H. Freeman and Company, New York.

Avdeef, A. (2014). Anomalous Solubility Behavior of Several Acidic Drugs. *ADMET & DMPK*, 33–42.

Bartolli, J., Turmo, E., Algueró, M., Boncompte, E., Vericat, M. L., Conte, L., Ramis, J., Merlos, M. and Forn, J. (1998). New azole antifungals. 2. Synthesis and antifungal activity of heterocycle carboxamide derivatives of 3-Amino-2-aryl-1-azolyl-2-butanol. *Journal of Medicinal Chemistry*, 1855-1868.

Bergström, C. A. S. and Avdeef, A. (2019). Perspectives in solubility measurement and interpretation. *ADMET & DMPK*, 88–105.

Bérubé, G. (2016). An overview of molecular hybrids in drug discovery. *Expert Opinion on Drug Discovery*, 281–305.

Blagden, N., de Matas, M., Gavan, P. T. and York, P. (2007). Crystal engineering of active pharmaceutical ingredients to improve solubility and dissolution rates. *Advanced Drug Delivery Reviews*, 617–630.

Bozorov, K., Zhao, J. and Aisa, A. (2019). 1,2,3-Triazole-containing hybrids as leads in medicinal chemistry: A recent overview, *Bioorganic & Medicinal Chemistry*, 3511-3531.

Brito, A. F., Moreira, L. K. S., Menegatti, R. and Costa, E. A. (2019). Piperazine derivatives with central pharmacological activity used as therapeutic tools, *Fundamental and Clinical Pharmacology*, 13-24.

Brittain, H. G. (1995). *Physical Characterization of Pharmaceutical Solids*, Marcel Dekker Inc., New York.

Buckley, S. T., Frank, K. J., Fricker, G. and Brandl, M. (2013). Biopharmaceutical classification of poorly soluble drugs with respect to "enabling formulations." *European Journal of Pharmaceutical Sciences*, 8–16.

Chai, X., Zhang, J., Hu, H., Yu S., Sun, Q., Dan, Z., Jiang Y. and Wu, Q. (2009). Design, synthesis, and biological evaluation of novel triazole derivatives as inhibitors of cytochrome P450 14a-demethylase, *European Journal of Medicinal Chemistry*, 1913–1920.

Correa, J. C. R., Reichman, C., Salgado, H. R. N. and Vianna-Soares, C. D. (2012). Performance characteristics of high performance liquid chromatography, first order derivative UV spectrophotometry and bioassay for fluconazole determination in capsules. *Quimica Nova*, 530–534.

Daina, A., Michielin, O. and Zoete, V. (2017). Swiss ADME: a free web tool to evaluate pharmacokinetics, drug-likeness and medicinal chemistry friendliness of small molecules. *Scientific Reports*, 42717.

Dearden, J. C. and Bresnen, G. M. (2005). Thermodynamics of Water-octanol and Water-cyclohexane Partitioning of some Aromatic Compounds. *International Journal of Molecular Sciences*, 119-129.

Delgado, D. R., Bahamón-Hernandez, O., Cerquera, N. E., Ortiz, C. P., Martínez, F., Rahimpour, E., Jouyban, A. and Acree, W. E. (2021). Solubility of sulfadiazine in (acetonitrile + methanol) mixtures: Determination, correlation, dissolution thermodynamics and preferential solvation. *Journal of Molecular Liquids*, 114979.

Egbuta, C., Lo, J. and Ghosh, D. (2014). Mechanism of inhibition of estrogen biosynthesis by azole fungicides. *Endocrinology*, 4622–4628.

El-Bayouki, K. and Basyouni, W. M. (2010). Thiazolopyrimidines without bridge head nitrogen: thiazolo[4,5-d]pyrimidines. *Journal of Sulfur Chemistry*, 551-590.

Emami, S., Ghobadi, E., Saednia, Sh. and Hashemi, S. M. (2019). Current advances of triazole alcohols derived from fluconazole: Design, in vitro and in silico studies, *European Journal of Medicinal Chemistry*, 173-194.

Fahmy, H., Sherif, A., Rostom, F., Saudi, M. N., Zjawiony, K. J. and Robins, D. J. (2003). Synthesis and in vitro evaluation of the anticancer activity of novel fluorinated thiazolo[4,5-d]pyrimidines. *Archiv der Pharmazie [Pharmacy Archives]*, 216–225.

Franks, F. (1993). Solid Aqueous Solutions. *Pure and Applied Chemistry*, 2527-2537.

Fromtling, R. A., Castaer, J. and Serradell, M. N. (1985). Fluconazole, *Drugs of the Future*, 982.

Gewald, K. (1966). Reaktion von methylenaktiven nitrilen mit senfolen und schwefel. *Journal für praktische Chemie [Journal of Practical Chemistry]*, 26–30.

Guillon, R., Pagniez, F., Picot, C., Hédou, D., Tonnerre, A., Chosson E., Duflos, M., Besson, T., Logé, C. and Le Pape, P. (2013) Discovery of a novel broad-spectrum antifungal agent, derived from albaconazole, *ACS Medicinal Chemistry Letters*, 288-292

Heeres, J., Meerpoel, L. and Levi, P. (2010). Conazoles. *Molecules*, 4129–4188.

Hsu, J. L., Ruoss, S. J., Bower, N. D., Lin, M., Holodniy, M. and Stevens, D. A. (2011). Diagnosing invasive fungal disease in critically ill patients. *Critical. Reviews in Microbiology*, 277-312.

Jouyban, A., Acree, W. E. and Martínez, F. (2020). Dissolution thermodynamics and preferential solvation of ketoconazole in some {ethanol (1) + water (2)} mixtures. *Journal of Molecular Liquids*, 113579.

Kalepu, S. and Nekkanti, V. (2015). Insoluble Drug Delivery Strategies: Review of Recent Advances and Business Prospects. *Acta Pharmaceutica Sinica B*, 442-453.

Kansy, M., Avdeef, A. and Fischer, H. (2004). Advances in screening for membrane permeability: high-resolution PAMPA for medicinal chemistsAuthor links open overlay panel, *Drug Discovery Today: Technologies*, 349-355.

Kempinska, D., Chmiel, T., Kot-Wasika, A., Mrózb, A., Mazerska, Z. and Namiesnik J. (2019). State of the art and prospects of methods for determination of lipophilicity of chemical compounds. *TrAC Trends in Analytical Chemistry*, 54-73.

Kerns, E., Di, L. (2008). *Drug like Properties: Concepts, Structure Design and Methods*, Academic Press, New York.

Kuppast, B. and Fahmy, H. (2016) Thiazolo[4,5-d]pyrimidines as a privileged scaffold in drug discovery. *European Journal of Medicinal Chemistry*, 198-213.

Noorizadeh, H., Sajjadifar, S. and Sobhanardakani, S. (2014). Prediction of octanol–water partition coefficients of organic chemicals by QSAR models. *Toxicological and Environmental Chemistry*, 1267–1278.

Ombrato, R. et al. Preparation of fused heterocyclic compounds as antibacterial agents. PCT, 2017211760, 20174.

Perlin, D. S., Shor, E. and Zhao, Y. (2015). Update on antifungal drug resistance, *Current Clinical Microbiology. Reports*, 84–95.

Raevsky, O. A., Grigor'ev, J. and Trepalin, S. V. (1999). *HYBOT program package*, RU 990090.

Rathi, A. K., Syed, R., Shin, H. S. and Patel, R. V. (2016). Piperazine derivatives for therapeutic use: a patent review (2010-present), *Expert Opinion on Therapeutic Patents*, 777–797.

Sangster, J. (1997). *Octanol-Water Partition Coeffcients: Fundamentals and Physical Chemistry*, John Wiley & Sons, Chichester.

Shah, P., Goodyear, B. and Michniak-Kohn, B. B. (2017). A review: Enhancement of solubility and oral bioavailability of poorly soluble drugs. *Pharmaceutical Journal*, 161-173.

Sheng, C., Wang, W., Che, X., Dong, G., Wang, S., Ji, H., Miao, Z., Yao, J. and Zhang, W. (2010). Improved model of lanosterol 14a-demethylase by ligand-supported homology modeling: validation by virtual screening and azole optimization. *Chem Med Chem*, 390–397.

Slavin, M. A. and Thursky, K. A. (2016) Isavuconazole: a role for the newest broad-spectrum triazole. *Lancet*, 726–728.

Xie, Y., Han, S., Shi, H., Du, C. and Zhao, H. (2016). Measurement and modelling of econazole nitrate in twelve pure organic solvents at temperatures from 278.15 K to 318.15 K. *Journal of Chemical Thermodynamics*, 59-68.

Xu, R., Zheng, M., Farajtabar, A. and Zhao, H. (2018). Solubility modelling and preferential solvation of adenine in solvent mixtures of (N,N-dimethylformamide, N-methyl pyrrolidone, propylene glycol and dimethyl sulfoxide) plus water. *Journal of Chemical Thermodynamics*, 225–234.

Zhou, C. H. and Wang, S. (2012). Recent researches in triazole compounds as medicinal drugs, *Current Medicinal Chemistry*, 239-280.

Zirngibl, L. (1998). *Antifungal azoles: a comprehensive survey of their structures and properties,* Weinheim, New York.

Chapter 2

Pyrimidine Ring Containing Natural Products and their Biological Importance

Ravindra K. Rawal[1,2,*]
Bishal Dutta[1]
and Pitambar Patel[1]

[1]Natural Product Chemistry Group, Chemical Sciences and Technology Division, CSIR-North East Institute of Science and Technology, Jorhat, Assam
[2]Academy of Scientific and Innovative Research (AcSIR), Ghaziabad, Uttar Pradesh, India

Abstract

Natural products remain an important source as well as inspiration for modern drug discovery. Among the natural products, heterocyclic compounds are at the forefront of drug discovery because of their high structural diversity, vast biological activities, and toxicity. In the family of heterocyclic natural products, the alkaloid class of natural products is the crucial group of compounds that are widely used in medicinal chemistry and drug discovery. However, this book chapter is mainly focused on the pyrimidine ring containing natural products and their biological activities along with clinical and *in vitro* applications. Substituted pyrimidine ring is widely found in many bioactive natural products such as vitamins, health supplements, antiviral, antifungal, antibacterial, etc. Based on their structural features, the pyrimidine class of natural products is divided into five main groups namely; i) tethered pyrimidines, ii) fused pyrimidines, iii) pyrimidine nucleoside and nucleotides and iv) vitamins and v) toxins. Each group is further divided

---

* Corresponding Author's Email: rawal.ravindra@gmail.com; drrkrawal@neist.res.in.

In: Pyrimidines and Their Importance
Editor: Roger G. Ward
ISBN: 979-8-88697-656-4
© 2023 Nova Science Publishers, Inc.

into sub-groups based on the nature of substituents and their relative biological activities are discussed in detail.

Keywords: pyrimidine, purine, alkaloids, antibiotics, vitamins, natural products

1. Introduction

Natural products have been utilized in folklore for the treatment of various ailments since ancient time yet their importance still persists in modern drug discovery. Natural products can be used as medicine in the form of mixtures or concentrated plant extracts without the need for the isolation of active compounds. In contrast, modern medicine needs the purification and isolation of one or two active phytochemicals, which sometimes makes them ineffective. Furthermore to the discovery of novel chemical compounds for medicinal utilization, natural products serve as a key basis as prospective lead phytochemicals for the discovery of new and more effective drug candidates through structural alteration. Even though natural products exhibit diverse and complicated chemical structures, plant secondary metabolites appear to be more biologically friendly and drug-like than those obtained from exclusively synthetic sources. As a result, natural compounds are thought to be superior prospects for future medication development. Natural products and their derivatives possess a wide spectrum of pharmacological activities and are utilized for the treatment of the most prevalent human diseases, comprising infectious diseases, cancers, diabetes, digestive system-related diseases, and cardiovascular system-related diseases. The annual world-wide market of medicine is either from natural products or related drugs (natural product derived, semisynthetic and natural product mimics, etc) is about 35%. Moreover, all the new chemical entities that entered into the market in the timespan of 1981-2002 from natural sources are about 28%. In the last four decades, about 48% of approved antibacterial agents are either from natural products or natural product derivatives.

Natural products have been the utmost prosperous source of potential drug leads and drug candidates due to their unique structural diversity. Among the natural products, heterocyclic compounds play a vital role in drug discovery. In the drug discovery process, heterocycles are the major structural units in the marketed drugs as well as in the targets of medicinal chemistry. Heterocycles can be used to alter the polarity, hydrogen bonding, and lipophilicity of molecules, which improves the physicochemical, pharmacological, pharmacokinetic, and toxicological aspects of drug candidates and, eventually, finished pharmaceuticals.

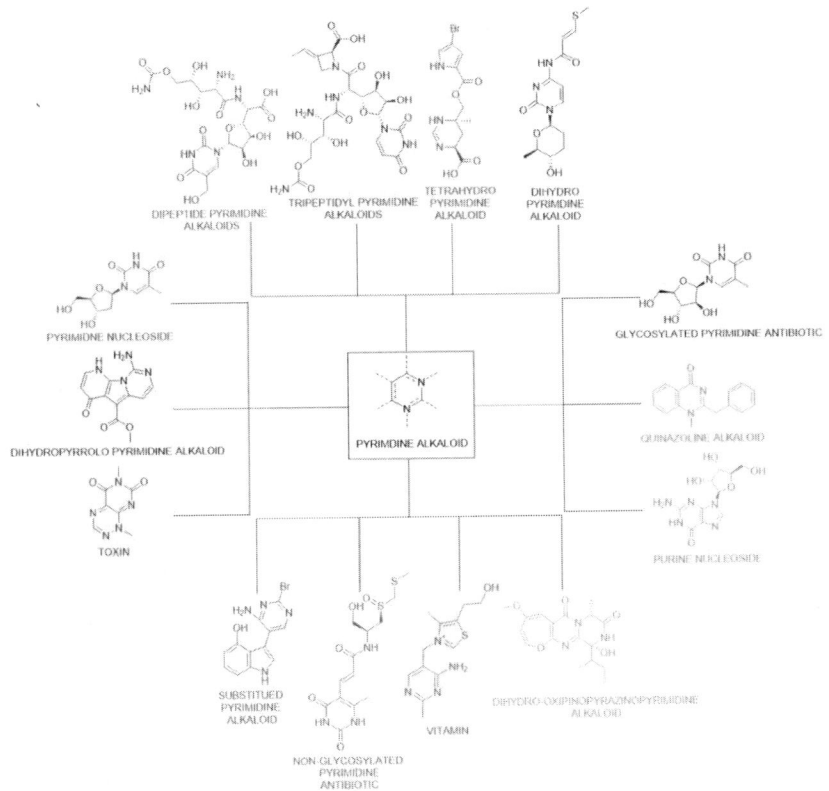

Figure 1. Pyrimidine ring containing natural products from different classes.

As a major class of heterocyclic natural products, alkaloids have been used globally as a source of medicines to treat various illnesses and diseases. In this book chapter, alkaloids containing pyrimidine rings (Figure 1) and their biological activities are discussed. Pyrimidine-containing alkaloids are extensively found in vitamins, antibiotics, antiviral, antibacterial, antifungal, anticancer compounds, and various health supplements as depicted in Figure 2. Drugs such as Edoxudine is an antiviral alkaloid containing pyrimidine ring, Acyclovir, a common antiviral medicine used to treat herpes, and Abacavir, an anti-HIV drug, have structural similarities to guanosine, a pyrimidine alkaloid. The pyrimidine alkaloids in this chapter are meticulously categorized according to their structures and/or biological activities. The book chapter will provide a summary of naturally occurring pyrimidine alkaloids and their biological activities, with an emphasis on the significance of this heterocyclic scaffold in the natural products synthesis.

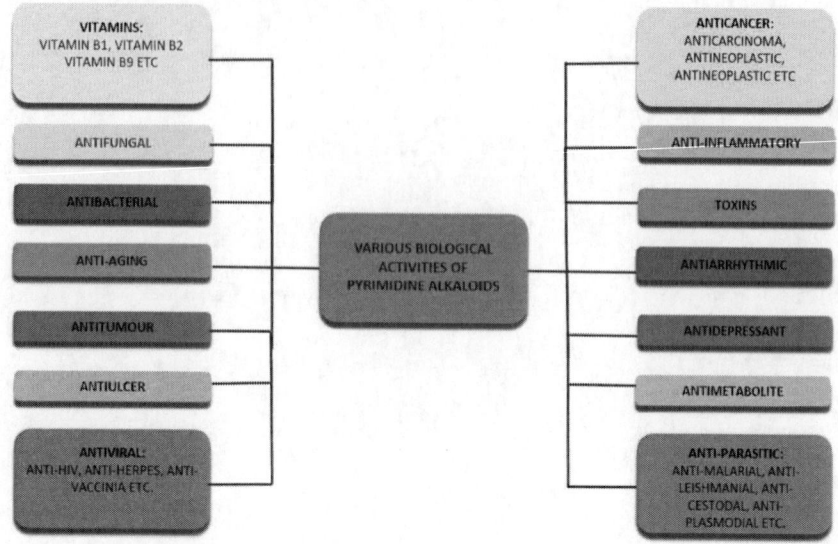

Figure 2. Various biological activities of pyrimidine ring containing alkaloids.

2. Tethered Pyrimidine Alkaloids

2.1. Substituted-Pyrimidine Alkaloids

Psammopemmins (**1-3**) are alkaloids of the variolin type which are isolated from Antarctic marine sponge *Psammopemma sp.* in the form of amine salts as shown in Figure 3 (Pauletti et al., 2010). They contain a novel 4'-amino-2'-bromopyrimidine-5-yl scaffold, which is hitherto unknown in the natural products chemistry. Psammopemmins (**1-3**) are not active as such but their semisynthetic analogues showed central nervous system (CNS) and antimalarial activity. Psammopemmin A (**1**) binding affinity was evaluated toward eleven 5-hydroxytryptamine (5-HT) receptor subtypes. It is active against 5-HT$_{1D}$ and D$_5$ at >10μM, and plasmodium falciparum at >33μM while it does not show any cytotoxic activity. During the secondary screening, it did not show binding efficacy of [3H]lysergic acid diethylamide (LSD) to 5-HT$_{1D}$.

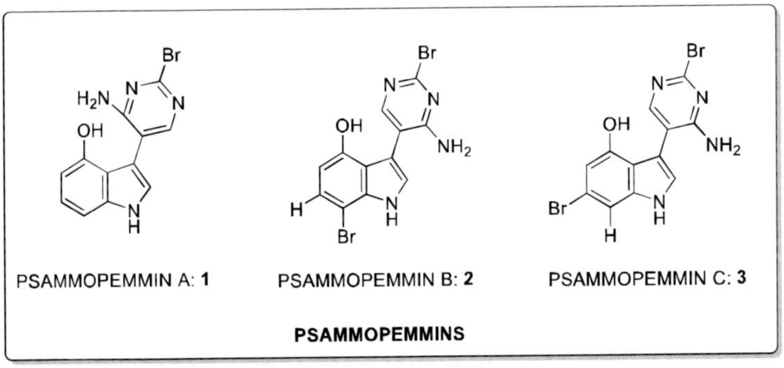

PSAMMOPEMMINS

PSAMMOPEMMIN A: **1**   PSAMMOPEMMIN B: **2**   PSAMMOPEMMIN C: **3**

Figure 3. Psammopemmin alkaloids.

Meridianins (**4-10**) are another set of variolin-type alkaloids and are isolated from *Aplidium meridianum* (Figure 4). They are also regarded as guanidine-based alkaloids since they include a guanidine moiety disguised as a 2-aminopyrimidine ring (Gompel et al., 2004). Meridianin A (**4**) meaningfully inhibited the binding of radioligand LSD with 5-HT2B (Ki = 150 nM). These preliminary findings indicate meridianin A (**4**) is a marginally more selective inhibitor of serotonin receptor subtypes. It displays activity in primary and secondary DAT screening (Ki = 2.35 μM). The significant antimalarial activity and cytotoxicity of meridianin A (**4**) were screened and found most potent (IC$_{50}$ = 12 μM) and also exhibited mild antimalarial activity. Meridianin A (**4**) also demonstrated cytotoxicity against A549 lung cancer cells at approximately equimolar concentrations to its antimalarial activity (IC$_{50}$ = 15 μM). This was unexpected because cytotoxic effects of meridianin A (**4**) in HepG2, HT29, LMM3 and SH-SY5Y cell lines (IC50 >100 μM) were not detected, earlier. Meridianin D (**7**) had limited anticancer activity against of several tumor cell lines, however, it was cytotoxic to LMM3 (murine mamary adenocarcinoma cell line). Except for meridianin G (**10**), all meridianins, particularly meridianin B (**5**) and E (**8**) were discovered to inhibit protein kinases such as Cyclin-dependent kinases (CDKs), glycogen synthase kinase-3 (GSK-3), protein kinase A (PKA), etc. in the micromolar and sub-micromolar range (Dong et al., 2020). Meridianins (**4-10**) can infiltrate cells and affect the activation of kinases, which are crucial for cell division and death. As a result, they can inhibit cell proliferation and trigger apoptosis also.

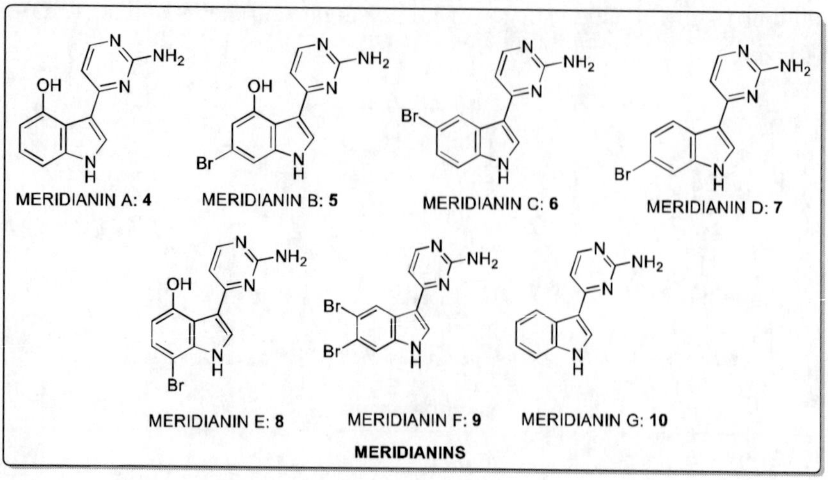

Figure 4. Meridianin alkaloids.

Heterostemma brownii Hayata (Asclepiadaceae), a plant used in Taipei traditional medicine to cure tumors, is a source of heteromine A–H (**11-18**) that may be extracted from its aerial portions (Lin et al., 1997). Heteromines (**11-18**) have demonstrated cytotoxicity in a variety of cancer cell lines, including hepatoma, lymphoma, esophageal carcinoma, and leukemia cells (Roggen et al., 2009). However, these substances don't appear to have more inhibition against different microorganisms, and their cytotoxicity seems to be restricted to cancer cell lines.

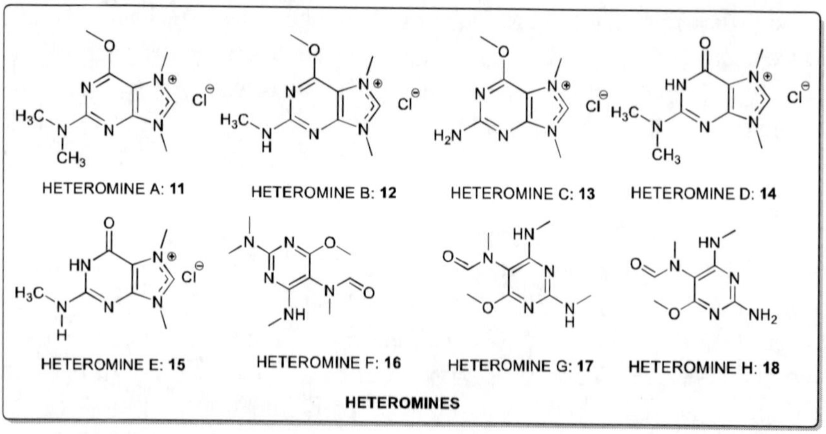

Figure 5. Heteromines alkaloids.

Cytosine (**19**) is the nucleobase found in both deoxyribonucleic acid (DNA) and ribonucleic acid (RNA) that is utilised to store and transfer genetic information inside of a cell. It is the precursor of cytosine nucleotides and therefore of ribo- and deoxyribonucleic acids. The DNA nucleotide cytosine can be found in the methylated form of 5-methylcytosine (**20**). Between species, 5-methylcytosine (**20**) has quite different functions. In bacteria, it is widely used as a marker to prevent DNA from being cut by native restriction enzymes that are methylation-sensitive. It appears in the CpG, CpHpG, and CpHpH sequences of plant DNA (where H is A, C, or T). The 5-Methylcytosine is mostly found in CpG dinucleotides in plants, animals, and fungi (Johnson and Coghill, 1925). Archaeal and eukaryotic tRNAs frequently include 5-methylcytosine (**20**) in their anticodon loop and central core, where it contributes to the structural and metabolic stability of the molecule (Squires and Preiss, 2010). Numerous investigations have shown that certain Drosophila tRNAs that contain 5-methylcytosine (**20**) at position 38 are resistant to oxidative stress, heat shock, and ribonuclease-induced tRNA breakage.

The 5-(hydroxymethyl)-cytosine (**21**) is a DNA pyrimidine nitrogen base derived from cytosine. The hydroxymethyl group on the cytosine may be able to switch a gene on and off, making it potentially significant in the field of epigenetics. Initially identified in bacteriophages, (Wyatt and Cohen, 1952) 5-(hydroxymethyl)-cytosine (**21**) was subsequently revealed to be widely distributed in human and animal brains, as well as embryonic stem cells (Kriaucionis and Heintz, 2009). The 5-(hydroxymethyl)-cytosine (**21**) appears to be present in greatest concentrations in neuronal cells of the central nervous system. It is thought that this nitrogen base controls gene expression or triggers DNA demethylation (Guo et al., 2011; Warren, 1980).

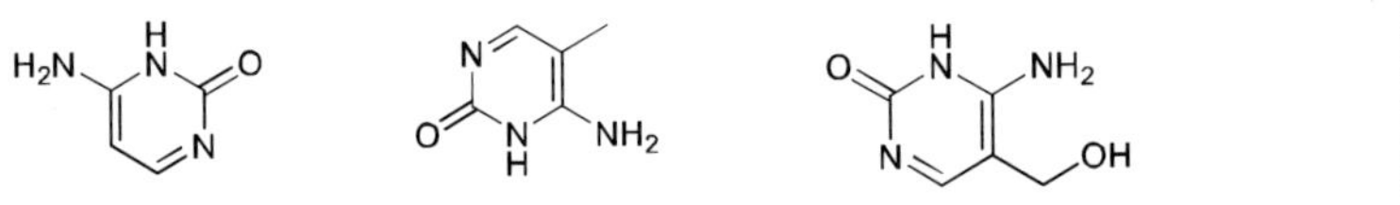

CYTOSINE: 19   5-METHYLCYTOSINE: 20   5-(HYDROXYMETHYL)-CYTOSINE: 21

CYTOSINES

Figure 6. Cytosines alkaloids.

Willardiine (**22**) was primarily isolated from the seeds of *Acacia willardiana*, it is a non-proteinogenic *L*-amino acid (Figure 7). One of the first chemicals whose pharmacokinetic characteristics were considerably altered by small changes in molecular structure was willardiine (**22**). It is a partial agonist of Ionotropic glutamate receptors. Willardiine (**22**) binds to α-amino-3-hydroxy-5-methyl-4-isoxazolepropionic acid (AMPA) and Kainate receptors in hippocampus or cortical neurons to stimulate the CNS. It has been utilized to elucidate the structural effects of activating AMPA/Kainate receptors by various agonists or antagonists due to its various binding affinities for each receptor (Jane et al., 1997).

Lathyrine (**23**) is a non-proteinogenic *L*-α-amino acid, extracted from *Lathyrus tingitanus*. It has the structure β-(2-aminopyrimidin-4-yl)alanine as depicted in Figure 7. Lathyrine (**23**) is a cytokinin in certain but not all plant cells, despite having antibacterial properties (Kafer et al., 2004). Zeatin (**24**; Figure 7) is a plant growth hormone generated from adenine which belongs to the cytokinin family (Mok and Mok, 1994). It was found in immature corn kernels of the genus *Zea*. Many plant extracts include zeatin (**24**) and its derivatives, which are also the main component of coconut milk. When zeatin (**24**) is sprayed on meristems, it encourages the formation of lateral buds and increases cell division to yield bushier plants. Zeatin riboside has an immunomodulatory effect by agonizing the mammalian adenosine A2A receptor. It also has numerous anti-aging benefits on human skin fibroblasts (Rattan and Sodagam, 2005). When paired with auxin, zeatin (**24**) is known to encourage callus initiation, delay vegetable yellowing, and stimulate the growth and flowering of auxiliary stems. It may be used to promote seed germination and seedling growth as well (Lappas, 2015).

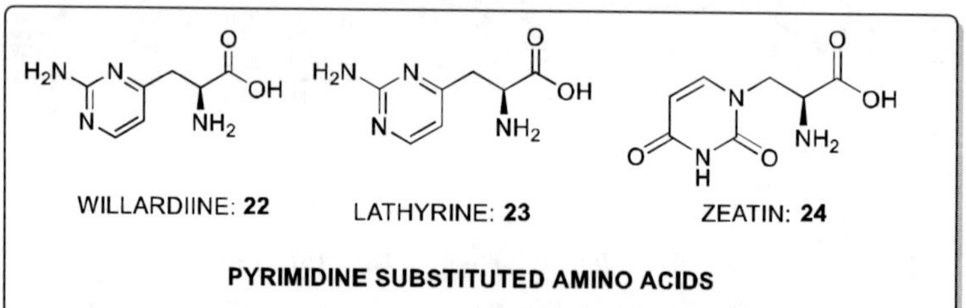

Figure 7. Pyrimidine substituted amino acids alkaloids.

Annomontine (**25**) is a pyrimidine-β-carboline alkaloid as shown in Figure 8. It is found in *Annona montana* and *Annona foteida* (Leboeuf et al., 1982). In mice, Animontine has been found to have a small amount of analgesic and anti-inflammatory action. It has also been demonstrated to have an antileishmanial impact when used *in vitro* against promastigotes of *Leishmania braziliensis* (Costa et al., 2006). The *N*-hydroxyannomontine (**26**; Figure 8) is an antileishmanial pyrimidine-β-carboline alkaloid found in *Annona foetida* (Costa et al., 2006). It possesses significant antitumor activity towards human hepatocellular carcinoma cell line (HepG2), as revealed by *in vitro* cytotoxicity assay (Costa et al., 2018). *In vitro* assay depict that *N*-hydroxyannomontine (**26**) is effective against promastigote forms of *Leishmania braziliensis* (Costa et al., 2006).

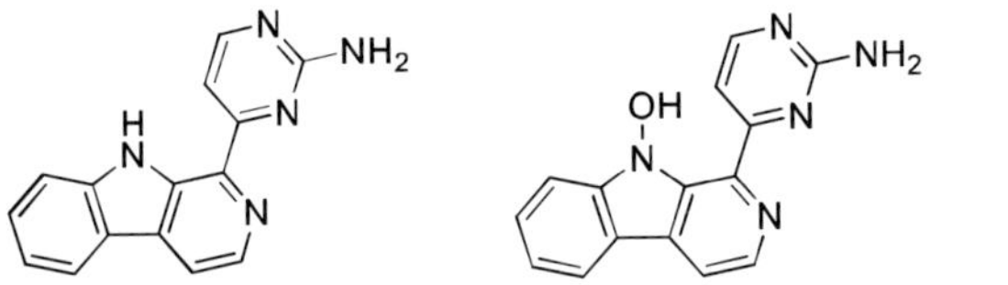

ANNOMONTINE: **25** N-HYDROXYANNOMONTINE: **26**

β-CARBOLINE ALKALOID

Figure 8. β-Carboline alkaloids.

Leucosolenamine A (**27**) and B (**28**) are isolated from *Leucosolenia*, a genus of calcareous sponges found in the Pacific Ocean as illustrated in Figure 9 (Ralifo et al., 2007). Leucosolenamine A (**27**) has a 2-aminoimidazole core skeleton that is replaced by *N,N*-dimethyl-5,6-diaminopyrimidine-2,4-dione at C-4 and benzyl group at positions C-5. Although the basic structure of leucosolenamine B (**28**) is the same, C-4 is changed to a 5,6-diamino-1,3-dimethyl-4-(methylamino)-3,4-dihydropyrimidin-2(1*H*)-one moiety (Roué et al., 2012). The chemistry of imidazole alkaloid has never seen a substitution pattern like this one before. Leucosolenamine B (**28**) does not affect on the mouse colon cancer C-38 cell line, but leucosolenamine A (**27**) has a modest cytotoxic effect.

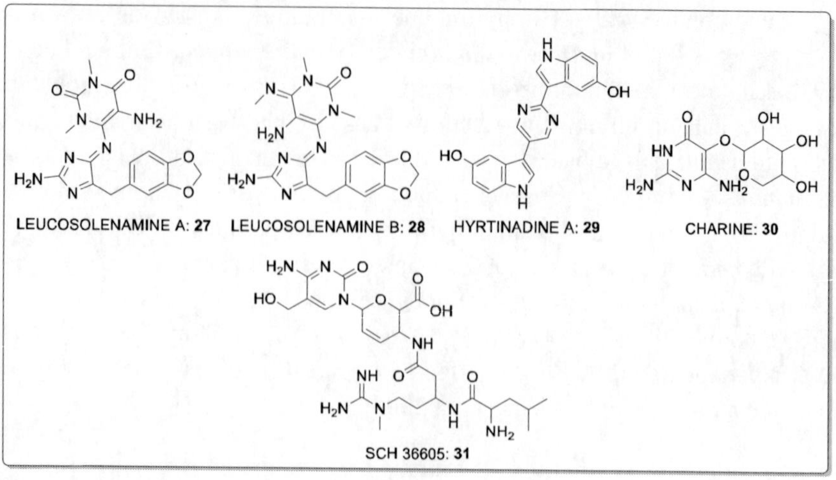

Figure 9. Miscellaneous pyrimidine substituted alkaloids.

The cytotoxic hyrtinadine A (**29**) was discovered in an Okinawan sea sponge called *Hyrtios sp.*, has a pyrimidine moiety and is the first known instance of a *bis*-indole alkaloid with 2,5-disubstituted pyrimidine ring in between (Figure 9). *In vitro* tests showed that hyrtinadine A (**29**) is cytotoxic to both human epidermoid carcinoma (KB cells) as well as murine leukaemia (L1210 cells). Charine (**30**) is a naturally occurring pyrimidine *O*-nucleoside and is a β-arabino nucleoside of divicine (Figure 9). It is found in *Momordica charantia*, a climbing vine used for various treatments in folk medicine. Its unripe fruits are mostly utilized for treating diabetes; the hypoglycemic effects have been demonstrated in both humans and animals (Raman and Lau, 1996). It is also used to treat gout, rheumatism, fevers, and stomachaches. It has been classified as a laxative, blood purifier, stimulant and antihelmintic in ayurvedic medicine (El-Gengaihi et al., 1995). SCH 36605 (**31**) is an anti-inflammatory compound isolated from the fermentation filtrate of a *Streptomyces sp* (Figure 9). It has anti-inflammatory properties in the adjuvant arthritic rat and the reverse passive Arthus response (Gullo et al., 1988).

2.2. Dihydro-Pyrimidine Alkaloids

Marine-derived *Streptomyces sp.* TPU1236A contains the antibiotics streptcytosine A–O (**32-46**) as depicted in Figure 10. Streptcytosine A (**32**), also known as Rocheicoside A, that belongs to the Amicetin group of antibiotics, whereas the streptcytosines B-E (**33-36**) are analogs of *de*-amosaminyl-cytosamine, 2,3,6-trideoxyglucopyranosyl cytosine. Being a potent antibiotic, it is found to be active against microorganisms, including eukarya, archaea, and bacteria species such as *Mycobacterium smegmatis* (Aksoy et al., 2016). Additionally, it demonstrates potential antimicrobial activity towards a variety of pathogens including *E. coli*, *C. albicans* and *MRSA* (Choudhary et al., 2017).

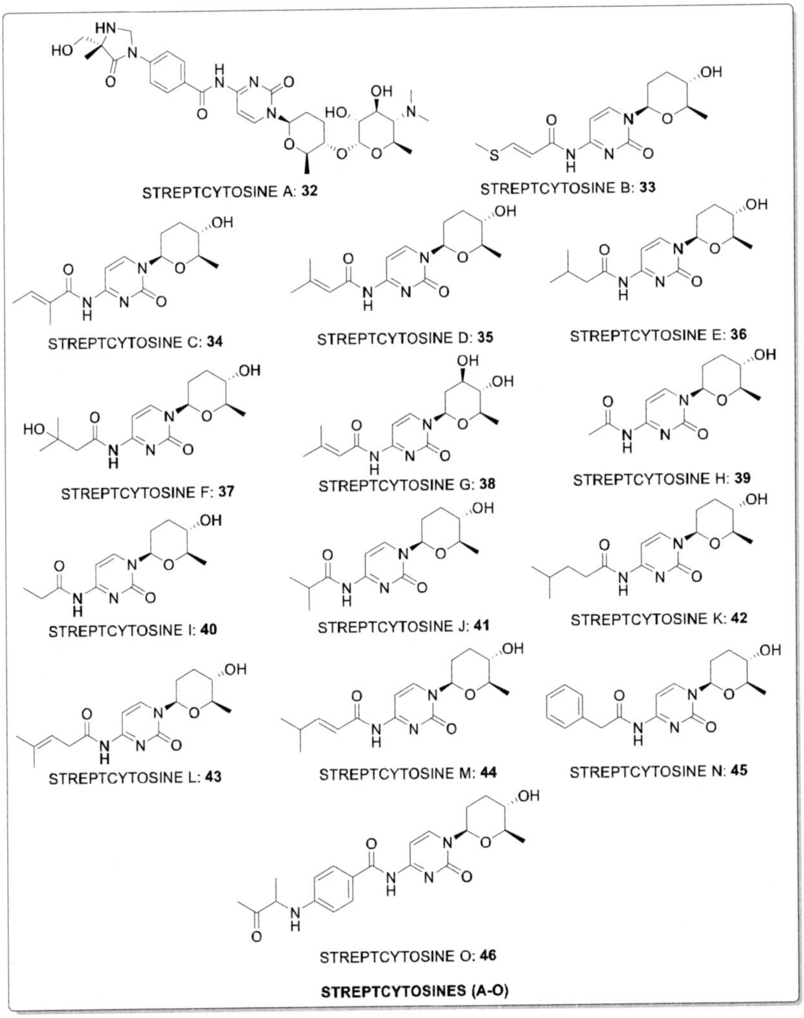

Figure 10. Streptcytosines (A-O) alkaloids.

Thymidine-5'-phosphate (**47**) or thymidine monophosphate is a nucleotide found in *Arabidopsis thaliana, Leishmania Mexicana* (Figure 11) and it is used in DNA as a monomer. It can be formed by the phosphorylation of thymidine *via* thymidine kinase catalysed pyrimidine salvage pathway. Thymidine monophosphate is a crucial metabolite needed for the precise replication of DNA genomes. Thymidine-3',5'-diphosphate (**48**) belongs to the class of pyrimidine deoxyribonucleoside 3',5'-biphosphates (Figure 11). It inhibits the activity of Staphylococcal Nuclease and Tudor Domain Containing 1 (SND1) in addition to being an anti-hepatocellular carcinoma (anti-HCC) agent. It also has *in vivo* anti-tumour activity.

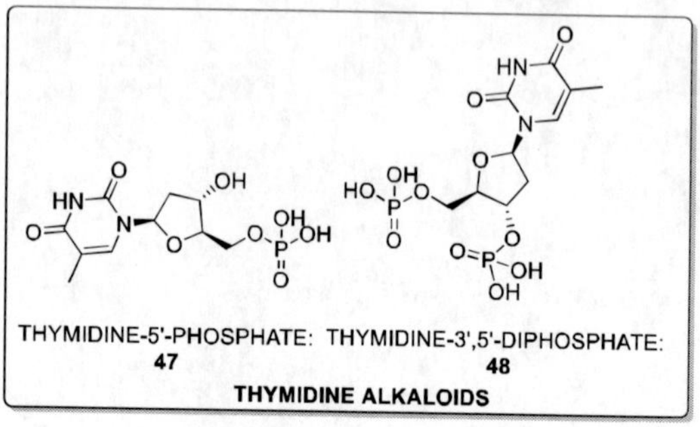

THYMIDINE-5'-PHOSPHATE: **47**
THYMIDINE-3',5'-DIPHOSPHATE: **48**
THYMIDINE ALKALOIDS

Figure 11. Thymidine alkaloids.

2.3. Tetrahydro-Pyrimidine Alkaloids

Manzacidin (**49-54**) is a family of alkaloids derived from marine sponges that have an ester-linked pyrrole and pyrimidine component as illustrated in Figure 12. The living fossil *Astrosclera willeyana* was discovered to possess manzacidin D (**52**), whereas the Okinawan sponge *Hymeniacidon sp.* yielded three bromopyrrole-containing derivatives manzacidin A–C (**49-51**) (Jahn et al., 1997).

Sparoxomycin A1(**55**) and A2 (**56**) are isolated from the mycelium and fermentation broth of *Streptomyces sparsogenes* SN-2325, which are pyrimidinylpropenamide antibiotics as shown in Figure 13. They act as inducers of the temperature-sensitive Rous Sarcoma Virus-induced flat reversion of NRK cells (Ubukata et al., 1996). These compounds closely resemble structurally with sparsomycin.

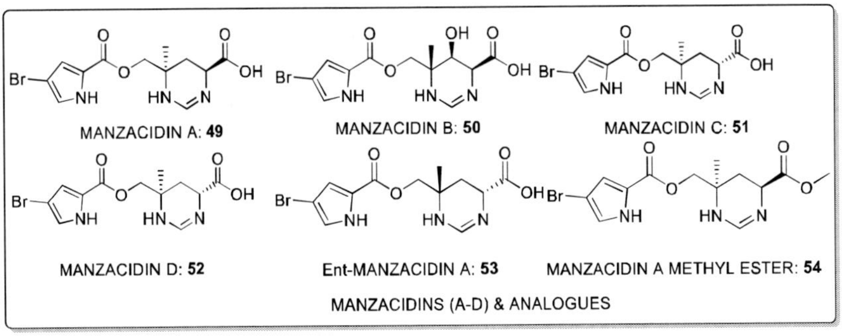

Figure 12. Manzacidins (A-D) alkaloids and their analogues.

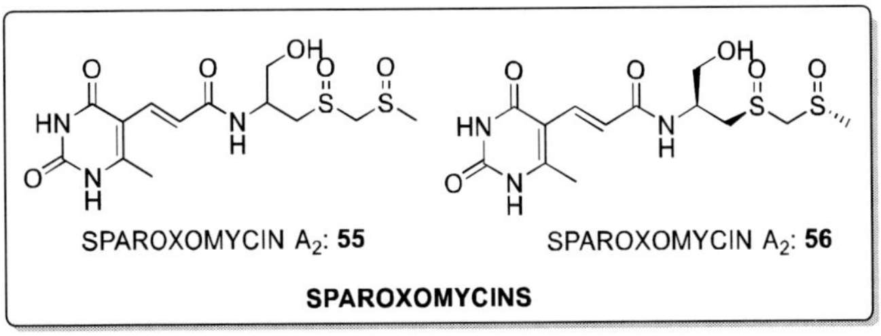

Figure 13. Sparoxomycins alkaloid.

3. Fused Pyrimidine Alkaloids

3.1. Quinazoline Alkaloids

Glycorine (**57**), glycosine (Arborine) (**58**), and glycosminine (**59**) are quinazoline alkaloids extracted from *Glycosmis Correa* as depicted in Figure 14. These alkaloids belong to the 4-quinazoline scaffold category. Glycosminine (**59**) shows moderate antifungal activity against *Candida albicans* (Xu et al., 2011). Arborine (**58**) was found to be a growth inhibitor of the larvae of *Drosophila melanogaster* (AhDlad et al., 1996). It also demonstrates formidable antibacterial activity against multidrug-resistant *Staphylococcus aureus* with the lowest MIC (minimum inhibitory concentration) value and MBC (minimum bactericidal concentration) value. When exposed to arborine (**58**), significant morphological changes were seen in multidrug-resistant *Staphylococcus aureus* such as uneven cell surfaces, cell shrinkage, and cell membrane damage (Murugan et al., 2020).

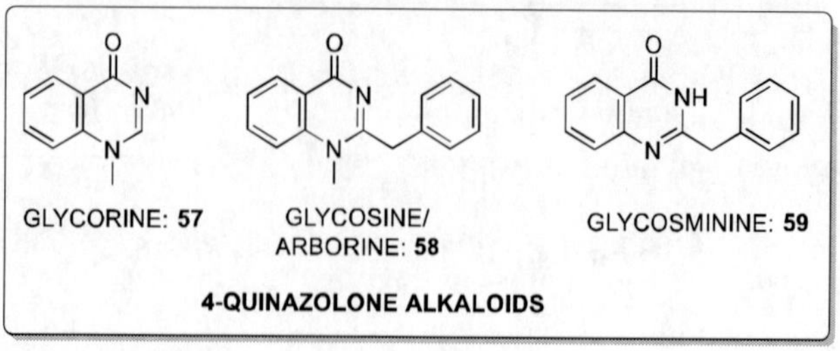

GLYCORINE: 57   GLYCOSINE/   GLYCOSMININE: 59
                   ARBORINE: 58

4-QUINAZOLONE ALKALOIDS

Figure 14. 4-Quinazolone alkaloids.

Febrifugine (**60**) and isofebrifugine (**61**) are two inter-convertible benzopyrimidine alkaloids that was isolated from *Dichroa febrifuga* and *Hydrangea umbellate* as shown in Figure 15 (Koepfli et al., 1949). Febrifugine (**60**) and isofebrifugine (**61**) have anti-malarial properties (Jiang et al., 2005), and the halogenated derivative of febrifugine (**60**), halofuginone is used in veterinary medicine as a coccidiostat. The antimalarial potency of febrifugine (**60**) is almost 100 times more potent than that of quinine against *P. lophurae* in duck models and >50 times more potent against *P. cynomolgi* infection in rhesus monkeys. Despite of having good antimalarial activity, they have been shown to possess significant toxicity in humans, preventing their clinical use. Febrifugine (**60**) is known to cause gastrointestinal tract irritation and emetic effects as side effect.

Glycosmicine (**62**) was isolated from *Glycosmis Correa* (Figure 16). The quinazoline alkaloid 7-bromoquinazoline-2,4(1H,3H)-dione (**63**) is extracted from *Pyura sacciformis* (Niwa et al., 1988; Pakrashi and Bhattacharyya, 1968; Pakrashi et al., 1963). It is the first naturally derived brominated quinazolinedione. 1-Methyl-3-(2-phenylethyl)quinazoline-2,4(1H,3H)-dione (**64**) is quinazoline alkaloid extracted from the extracts of the Mexican *Zanthoxylum arborescens* seed husk (Dreyer and Brenner, 1980).

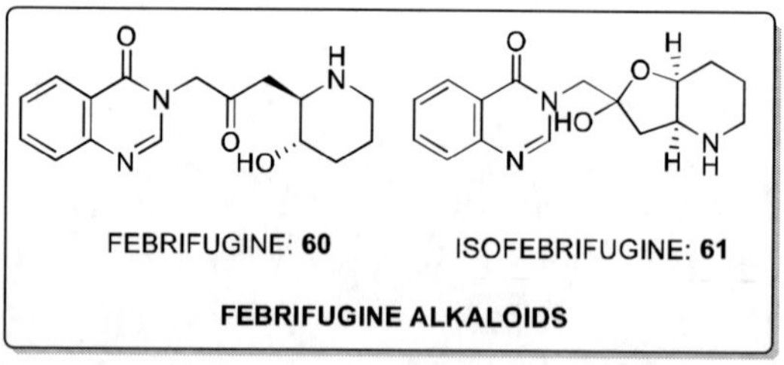

FEBRIFUGINE: 60   ISOFEBRIFUGINE: 61

FEBRIFUGINE ALKALOIDS

Figure 15. 4-Quinazolone-N3-substituted alkaloid.

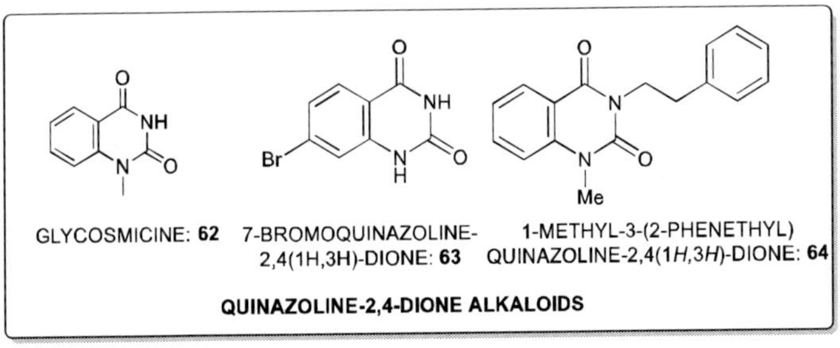

GLYCOSMICINE: 62   7-BROMOQUINAZOLINE-   1-METHYL-3-(2-PHENETHYL)
                    2,4(1H,3H)-DIONE: 63   QUINAZOLINE-2,4(1H,3H)-DIONE: 64

QUINAZOLINE-2,4-DIONE ALKALOIDS

Figure 16. Quinazoline-2,4-dione alkaloid.

Vasicine (**65**) or peganine is a quinazoline alkaloid (Soni et al., 2008) as depicted in Figure 17. It is found in *Justicia adhatoda* and also in *Peganum harmala*. Vasicine (**65**) has a cardiac-depressant effect and is also reported to have a uterine stimulant effect. It is generally used as an expectorant and bronchodilator. It also demonstrate oxytoxic as well as abortifacient properties. *S. aureus*, *S. epidermidis*, and *K. pneumoniae* were all susceptible to the antibacterial effects of vasicine (**65**) (Ignacimuthu and Shanmugam, 2010). Vasicine (**65**) has a stimulatory impact on the rat/guinea pig uterus and tracheal muscle in addition to other tissues, as was discovered during research of the uterine stimulant effect of vasicine (**65**) (Gupta et al., 1977). In a prepared isolated rat mammary strip, it enhanced the effects of oxytocin. It also shown smooth muscle stimulant action and is utilised for bronchodilation and abortion (Gupta et al., 1978). When administered during labour, as a respiratory stimulant vasicine (**65**) can counteract the respiratory side effects of narcotic analgesics. It is also helpful in reducing postpartum hemorrhage. On top of these, vasicine (**65**) is reported to possess anti-diabetic, anticestodal, antileishmanial, anti-helminthic and anti-ulcer activities (Moloudizargari et al., 2013). Glycophymoline (**66**) is a quainazoline alkaloid extracted from *Glycosmis Correa* (Figure 17). The Glycophymoline (**66**) enriched extracts showed antinociceptive activity in swiss albino mice.

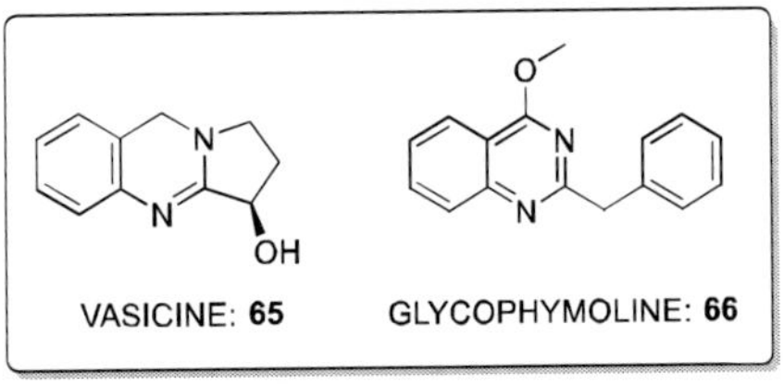

VASICINE: 65   GLYCOPHYMOLINE: 66

Figure 17. Miscellaneous quinazoline alkaloids.

3.2. Dihydropyrrolo-Pyrimidine Alkaloids

Variolins (**67-69**) are unconventional guanidine-based alkaloids with a wide range of potent biological effects wherein the guanidine moiety is present in the form of 2-aminoimidazol or 2-aminopyrimidine rings as shown in Figure 18. *Kirkpatrickia variolosa*, a rare and hard-to-access Antarctic sponge, is the source of these alkaloids (Trimurtulu et al., 1994). The earliest instances of pyrido[3',2':4,5]pyrrolo[1,2-*c*]pyrimidine-containing natural products are variolins (**67-69**) that are either terrestrial or marine. It has been demonstrated that variolin B (**68**) has strong pro-apoptotic properties. In addition to disrupting the cell cycle and inducing apoptosis, variolin B (**68**) inhibits colony formation in certain human cancer cell lines (Simone et al., 2005). It was also noticed to show antiviral activity against HSV-1 and Polio type I (Perry et al., 1994).

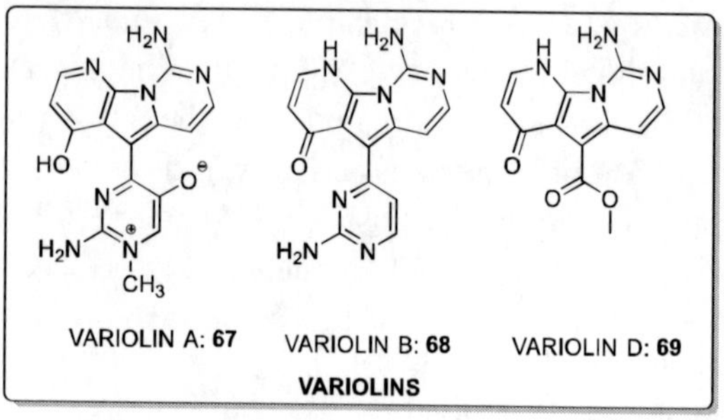

Figure 18. Variolin alkaloids.

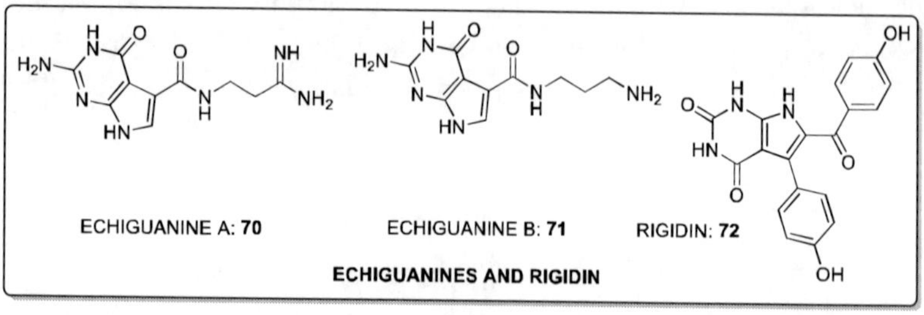

Figure 19. Echiguanine and rigidin alkaloids.

A strain of Streptomyces produces echiguanines (**70-71**), which are powerful inhibitors of phosphatidylinositol kinase (Nishioka et al., 1989) as shown in Figure 19. The strongest phosphatidylinositol kinase inhibitor yet identified, echiguanine A (**70**) is more effective than toyocamycin, 2,3-dihydroxybenzaldehyde, quercetin, or orobol. The Okinawan marine tunicate *Eudistoma cf. rigida* provided the pyrrolopyrimidine alkaloid rigidin (**72**; Figure 19), which has calmodulin antagonistic action (Kobayashi et al., 1990). A crucial structural component of many biologically active compounds is the presence of highly substituted pyrrole rings fused to six-membered rings. Natural antibacterial/antitumor chemicals and numerous bioactive nucleotide analogues, in particular, contain pyrrolo[2,3-*d*]pyrimidine ring systems, which calls for the development of new, more effective synthetic techniques.

3.3. Dihydro-Oxipinopyrazinopyrimidine Alkaloids

Oxepinamides (**73-79**) are derived from tripeptides containing anthranilyl that share a fused pyrimidinone moiety and an oxepin ring as depicted in Figure 20. They mostly exist in fungi eg. oxepinamide H (**76**) was isolated from *Aspergillus puniceus*. Some oxepinamides (**73-79**) have a strong affinity for liver X receptors (Liang et al., 2019; Lu et al., 2011) and may be used as drugs to treat diabetes, inflammation, atherosclerosis, and Alzheimer's disease. The marine-derived fungus *Aspergillus sp.* yields Protuboxepins (**80-81**), which are oxepin-containing alkaloids as shown in Figure 20 (Lee et al., 2011). Protuboxepins A and B (**80-81**) are among a rare group of naturally occurring substances with oxepin and diketopiperazine ring scaffold. In addition, they comprehend the relatively uncommon *D*-Phe residue. Protuboxepin A (**80**) exerts anticancer activity towards a number of cancer cell lines. It binds directly to α,β-tubulin and stabilizes tubulin polymerization, causing microtubule dynamics to be disturbed (Asami et al., 2012). This disturbance causes chromosomal misalignment and metaphase arrest by inducing apoptosis in cancer. Therefore, protuboxepin A (**80**) is a microtubule-stabilizing agent which is structurally different from formerly known microtubule inhibitors. Janoxepin (**82**) is an antiplasmodial oxepine-pyrimidinoen-ketopiperazine derived from *D*-leucine (Doveston et al., 2012). It was isolated from the *Aspergillus janus* fungus, which exhibits anti-plasmodial activity against the malaria parasite *Plasmodium falciparum* 3D7. Brevianamides O (**83**) and P (**84**) are isolated from *Aspergillus versicolor* fungus (Figure 20) (Borthwick, 2012). These compounds are 2,5-diketopiperazines, which are of indole class alkaloids as depicted from the chemical structures. These are significant secondary metabolite presents in *penicillium* spores and are responsible for inducing an inflammatory response in lung cells (Rand et al., 2005).

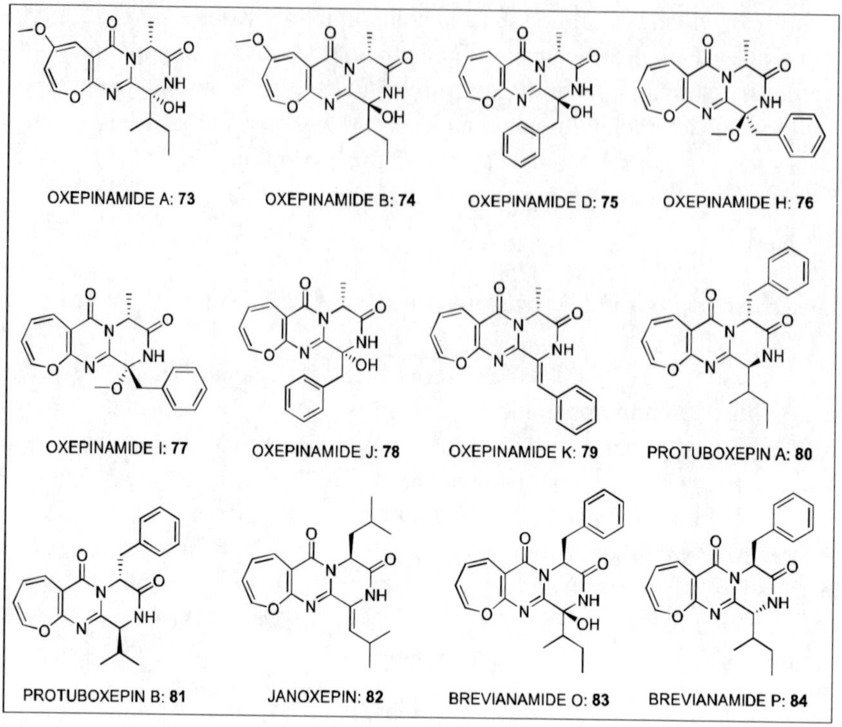

Figure 20. Dihydro-oxipinopyrazinopyrimidine alkaloids.

4. Nucleosides/Nucleotides

4.1. Pyrimidine Nucleosides/Nucleotides

2'-Deoxycytidine (**85**) is a pyrimidine 2'-deoxyribonucleoside containing cytosine as the nucleobase (Figure 21) (Kim et al., 2016). It is a natural product found in *Isodictya erinacea*, *Trypanosoma brucei* etc and also serves as a human, *Escherichia coli*, *Saccharomyces cerevisiae* and mouse metabolite, among other functions. It can be utilised as a precursor for the myelodysplastic syndromes (MDS) drug 5-aza-2'-deoxycytidine. 2'-deoxycytidine slows the cell cycle and prevents the development of MDS into leukaemia by interfering with the methylation of the P15/INK4B gene and boosting the production of P15/INK4B protein. The *N*(4)-hydroxycytidine (**86**) is a modified analog of cytidine. It induces mutations in RNA virions. The influenza (Flu), severe acute respiratory syndrome associated coronavirus (SARS-CoV), SARS-CoV-2, and Middle East respiratory syndrome coronavirus (MERS-CoV) are all susceptible to its broad spectrum antiviral action (Sticher et al., 2020). The *N*(4)-hydroxycytidine (**86**) serves as a human xenibiotic metabolite, a drug metabolite, an anti-corona virus agent, and an antiviral agent. It is orally bioavailable in mice but the bioavailability in non-human primates is poor. Cytidine (**87**) is a pyrimidine nucleoside comprising of a cytosine attached to ribose by a *β-N*1-glycosidic bond (Figure 21). Foods with a high RNA concentration, including organ meats and brewer's yeast, contain cytidine (**87**) (Jonasa et al., 2001). Additionally, it can be found in pyrimidine-rich foods like beer. It has attracted attention as a possible glutamatergic antidepressant medicine (Machado-Vieira et al., 2010) because, it has been discovered to influence neuronal-glial glutamate cycling, with supplementation lowering midfrontal/cerebral glutamate/glutamine levels (Yoon et al., 2009). Furthermore, cytidine (**87**) functions as a metabolite in humans, *Saccharomyces cerevisiae, Escherichia coli*, mice, and other organisms. Cytidine-5'-monophosphate (**88**) or cytidine monophosphate (CMP) is a pyrimidine nucleotide, utilised as a monomer in RNA (Figure 21) (Pascal, 2008). It is a phosphoric acid ester with the nucleoside cytidine. *Campylobacter jejuni* and *Mycoplasma gallisepticum* both naturally contain CMP. It functions as *Escherichia coli* in mouse and in humans as metabolite. The pyrimidine nucleoside triphosphate cytidine-5'-triphosphate (**89**), often known as cytidine triphosphate (CTP; Figure 21), is made up of three phosphate groups esterified to the deoxyribose sugar molecule (Buchanan et al., 2015). It works as a substrate for the synthesis of RNA. CTP (**89**) is found naturally in *Apis cerana* and *Homo sapiens*. Like ATP, CTP (**89**) is a high-energy molecule, but it only functions as an energy coupler in a small fraction of metabolic processes. In metabolic processes including the creation of glycerophospholipids, CTP (**89**) functions as a coenzyme for the activation and transfer of diacylglycerol and lipid head groups as well as the glycosylation of proteins (Blackburn et al., 2006). Moreover, it functions as an inhibitor of the pyrimidine biosynthetic enzyme aspartate carbamoyltransferase. Furthermore, CTP (**89**) serves as a mouse metabolite and an *Escherichia coli* metabolite. A crucial chemical needed for the formation of cell membranes is cytidine-5'-diphosphocholine (**90**), also known as CDP-choline or citicoline (Figure 21) (Adibhatla et al., 2002). Citicoline (**90**) is naturally occurring in the cells of human and tissue, particularly in the organs. It serves as an essential intermediary in the Kennedy pathway, which is used to synthesise phosphatidylcholine, a key phospholipid in the brain. A number of conditions, including cerebral ischemia, traumatic brain injury, learning and memory impairments, Alzheimer's and Parkinson's diseases, hypoxia, amblyopia, alcoholism, drug addiction, and glaucoma have been researched in relation to citicoline's potential therapeutic benefits. Since it is non-xenobiotic, citicoline (**90**) essentially has no negative effects. Citicoline (**90**) was originally studied for strokes. Few studies were successful in showing positive results while results from other studies have either failed to prove any benefit of citicoline (**90**) in strokes or showed ambiguous results. Clinical studies have shown that citicoline (**90**) enhances cognitive function in aged people. Cytidine monophosphate *N*-acetylneuraminic acid (**91**; Figure 21) is a nucleotide sugar used as a donor by glycosyltransferases for the synthesis of sugar chains. It is a natural product found in *Apis cerana* and *Homo sapiens*. It has a major role as metabolite in mice and humans.

Uridine (**92**) is a component of RNA, made up of uracil and D-ribose (Figure 21). It serves as a fundamental metabolite in humans and a metabolite for drugs. It is considered a non-essential nutrient since the human body can create it as needed, negating the need for supplementation in most cases. Because uridine (**92**) is essential to the brain's pyrimidine metabolism, it is the most common type of pyrimidine nucleosides taken up by the brain (Dobolyi et al., 2011). It provides the pyrimidine ring to nervous tissue and takes part in a variety of crucial metabolic pathways. Uridine (**92**) and its nucleotide derivatives may also contribute to the action of central nervous system as signaling molecules. The nervous system may be impacted by uridine (**92**) in a number of ways, too. Uridine (**92**) was found to be a potent component of sleep enhancing natural product isolated from the brainstem of sleep-deprived rats. The effect of uridine (**92**) was also studied on memory functions, neuronal plasticity, epilepsy models etc. 2'-Deoxyuridine (deoxyuridine) (**93**) is a pyrimidine 2'-deoxyribonucleoside having uracil as the nucleobase (Figure 21). It is present in all living organisms and can incorporate into DNA of both prokaryotic and eukaryotic cells. In the synthesis of thymidine, deoxyuridine acts as an intermediate. It functions as a metabolite in human, *Saccharomyces cerevisiae*, mouse and *Escherichia coli*. Deoxyuridine is a potentially toxic. Several foods contain it naturally outside the body, making it a useful indicator for diseases through consumption of those foods.

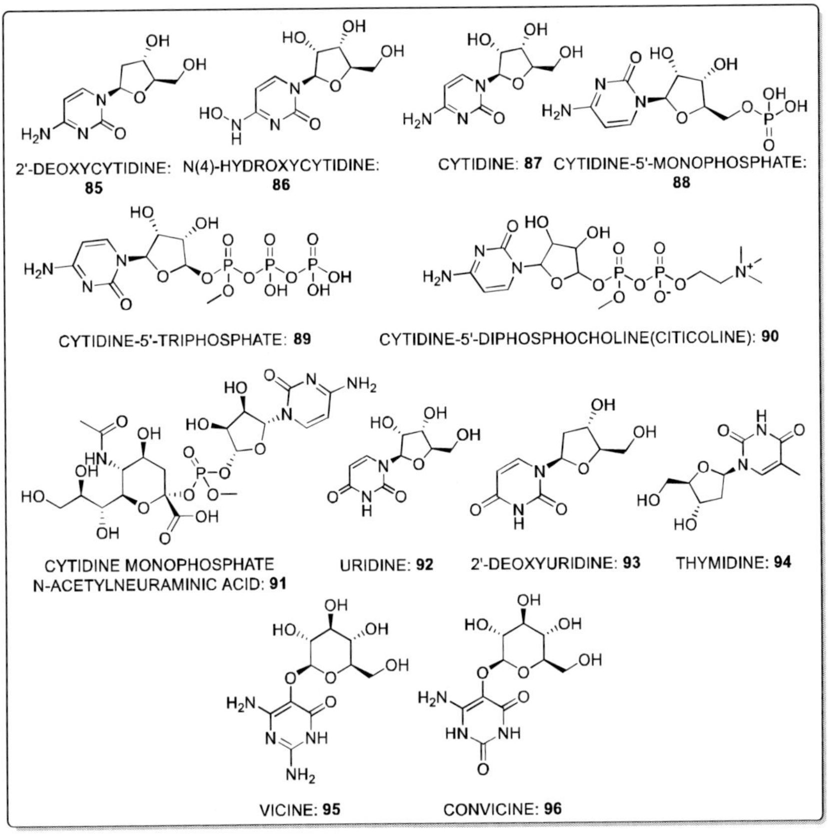

Figure 21. Pyrimidine Nucleosides/Nucleotides.

Thymidine (**94**) is a pyrimidine deoxynucleoside, made up of the sugar deoxyribose linked to the pyrimidine base thymine (Figure 21). All living organisms, including DNA viruses, contain the naturally occurring substance thymidine (**94**). It is also non-toxic and makes up one of the four nucleosides in DNA. Thymidine (**94**) pairs with Adenine as a constituent of DNA in the DNA double helix. It acts as metabolite in human, *Escherichia coli* and mouse. For various applications, different thymidine modified analogues are used such as azidothymidine is used in the treatment of HIV infection, tritiated thymidine is a radiolabeled thymidine used in cell proliferation assays, iododeoxyuridine is a radiosensitizer which enhances the amount of DNA damage caused by ionizing radiation, bromodeoxyuridine is used for the detection of proliferating cells, edoxudine is an antiviral drug etc.

Pyrimidine glycosides vicine (**95**) and convicine (**96**) are present in faba beans (Figure 21) as precursors of divicine and Isouramil, which are the primary factors in favism (Luzzatto and Arese, 2018); a potentially fatal genetic disorder that causes acute haemolysis after consuming faba beans (Rizzello et al., 2016). They mainly affect people having a hereditary lack of the enzyme glucose-6-phosphate dehydrogenase. Individuals with glucose-6-phosphate dehydrogenase lack are asymptomatic of haemolytic anaemia and for such a person an attack of acute haemolytic anaemia can appear out of nowhere and be very critical and life-threatening. Dark urine, abdominal pain, jaundice, and in most cases fever are symptoms of a sudden attack of favism.

4.2. Purine Nucleosides/Nucleotides

Guanosine (**97**) is a purine nucleoside required for metabolism, consists of guanine connected through a β-N9-glycosidic bond to a ribose ring as depicted in Figure 22 (Lewis, 2016). Guanosines can be present in pancreas, clover, coffee plant and in pollen of pines. It is necessary for an RNA splicing process in mRNA, in which a "self-splicing" intron disconnects itself from the mRNA message by cutting at both ends, re-ligating, and leaving just the exons on each side to be translated into protein. It shares structural similarities with acyclovir antiviral medicine, which is frequently utilized to treat herpes and the HIV infection. Deoxyguanosine (**98**) is a purine 2'-deoxynucleoside with the nucleobase guanine (Figure 22). Deoxyguanosine (**98**) is one of the four deoxyribonucleosides that make up DNA. It is found naturally in *Streptomyces antibioticus*, *Vitis vini fera* etc. It functions as a *Saccharomyces cerevisiae*, *Escherichia coli*, human, and mouse metabolite. It has long been known that deoxyguanosine (**98**) exerts strong cytotoxic effects on cultured mammalian cells. In human and murine T cells activated by mitogens, deoxyguanosine (**98**)-mediated inhibition of DNA synthesis has also been noted.

Adenosine (**99**) is a purine nucleoside base extensively distributed in nature in the guise of diverse derivatives (Figure 22). Adenosine (**99**) is one of the nucleoside building blocks of RNA (Mitchell and Lazarenko, 2008). Adenosine (**99**) derivatives contains the energy carriers adenosine monophosphate (AMP), adenosine diphosphate (ADP) and adenosine triphosphate (ATP; **101**) and used as therapeutic tool. Because of its vasodilatory properties, it can be employed in diagnostic myocardial perfusion stress imaging. Adenosine (**99**) has the ability to treat supraventricular tachycardia (SVT) because of its antiarrhythmic qualities (Goyal et al., 2019).

Adenosine (**99**) can be used to effectively end certain SVTs. Adenosine (**99**) can also terminate specific reentrant tachycardias that include the atrioventricular (AV) node, such as orthodromic AV reentrant tachycardia (AVRT), AV nodal reentrant tachycardia (AVNRT), and antidromic AVRT. It is also utilised as an addition to thallium or technetium myocardial perfusion scintigraphy in patients who are not able to endure sufficient stress testing with exercise (O'Keefe Jr et al., 1992). Adenosine analogue (NITD008) has been shown to directly block the dengue virus's recombinant RNA-dependent RNA polymerase (RdRp) by terminating the synthesis of its RNA chains which in turn inhibits peak viremia, increase in cytokines, and reduces death in infected animals, and also suggesting a potential novel flavivirus treatment. The hepatitis C virus (HCV) has been shown to be inhibited by 7-deaza-adenosine, another adenosine analogue. Ebola and Marburg viruses are resistant to the analogue BCX4430. Adenosine has also been demonstrated to have anti-inflammatory, hair-thickening, and sleep-inducing qualities (Oura et al., 2008). The toxicity of Adenosine (**99**) is kept to a minimum because of its short half-life. Although there have been reports of serious side-effects of adenosine use that includes prolonged asystole, heart block development and cardiac ischemia. 2'-Deoxyadenosine (**100**) is a purine nucleoside component of DNA (Figure 22). It is a natural product found in *Streptomyces jiujiangensis, Fritillaria przewalskii* etc. Both adenosine (**99**) and 2'-deoxyadenosine (**100**) are intermediates in the pathway of purine nucleotide degradation. Adenosine triphosphate (ATP) (**101**), an adenine nucleotide made up of three phosphate groups esterified to the 5-hydroxy methyl group of sugar moiety, supplies energy for a variety of cellular functions (Hassett et al., 1982). Condensate dissolution, nerve impulse transmission, muscular contraction, and chemical synthesis are a few examples of these processes. ATP (**101**) is often referred to as the power station of the cell and is found in all life forms. It contributes to the transportation of chemical energy throughout metabolic processes. ATP (**101**) serves as a micronutrient, a basic metabolite, and a nutraceutical. It serves as a coenzyme and is a precursor of DNA and RNA. Additionally, it is synthesised in cancer research to look at how it may be used to stop muscle atrophy and increase strength. Amino acid activation during protein synthesis, DNA and RNA synthesis, extracellular signaling, neurotransmission, and intracellular signaling are just a few of the biochemical processes that ATP (**101**) is involved in it (Scheeff and Bourne, 2005). Clinical studies have shown that ATP (**101**) reduces acute perioperative pain (Hayashida et al., 2005). Additionally, it has been proved to be a reliable and useful pulmonary vasodilator in people with pulmonary hypertension (Agteresch et al., 1999).

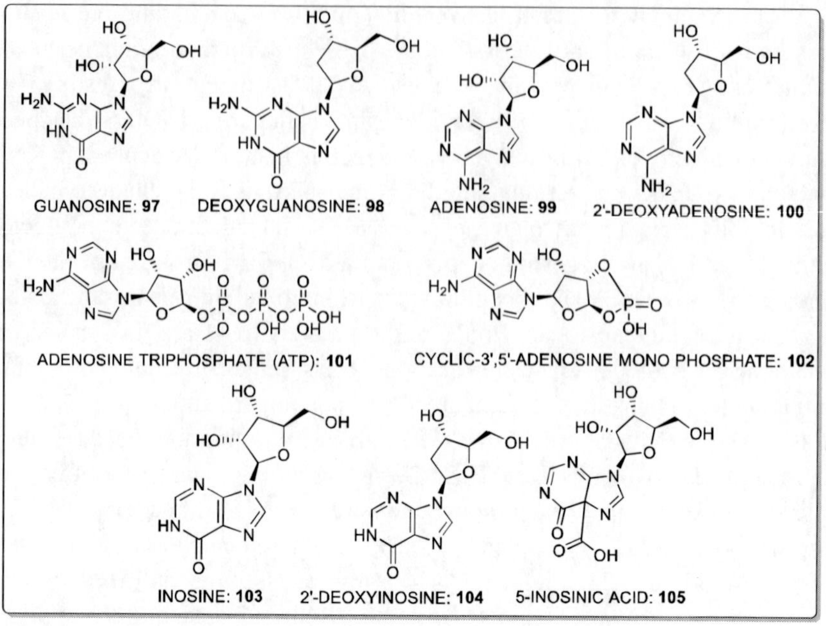

Figure 22. Purine Nucleosides/Nucleotides.

The adenine nucleotide cyclic 3', 5'-adenosine monophosphate (cAMP; **102**) has one phosphate group and is esterified to both the 3' and 5'-hydroxy group of the sugar moiety as shown in Figure 22 (Bos, 2006). The cAMP (**102**) serves as a mediator of action for several hormones, such as epinephrine, glucagon, and ACTH, and is a second messenger and an important intracellular regulator. The cAMP (**102**) is a derived from ATP (**101**), used by a wide range of organisms for intracellular signal transduction. Activating protein kinases is another function it plays. Additionally, cyclic AMP (**102**) binds to and controls the activity of cyclic nucleotide-binding proteins such as RAPGEF2 and Epac1 as well as HCN ion channels. Cyclic AMP (**102**) and kinases play a vital role in the control of glycogen, sugar, and lipid metabolism, among other metabolic activities. Recent research reveals that via controlling HCN, cyclic AMP (**102**) influences the way higher-order thinking functions in the prefrontal cortex. Another study contends that the development of various cancers is associated with the deregulation of cyclic

AMP (**102**) pathways and the irregular activation of cyclic AMP-controlled genes.

Inosine (**103**) is a naturally occurring purine nucleoside consist of a hypoxanthine connected to a ribose ring by a β-N_9-glycosidic bond (Figure 22). It frequently occurs in human transfer RNAs and is necessary for accurate translation of the wobble base pairs genetic code. Red meat, pork, and poultry all naturally contain inosine (**103**). Due to its ability to cause axonal rewiring, inosine (**103**) has been suggested as a management for spinal cord injury and as a post-stroke medication (Chen et al., 2002; Liu et al., 2006). Animal studies have recommended that inosine (**103**) has neuroprotective properties. Inosine (**103**) is metabolised into uric acid after ingestion, which may have advantages for multiple sclerosis patients as a natural antioxidant and peroxynitrile scavenger (MS) (Koch and De Keyser, 2006). Additionally, it has been discovered that in some species of farmed fish, either by itself or in conjunction with specific amino acids, inosine serves as a significant feed stimulant. Despite the lack of any scientific proof that it enhances muscle development, it continues to be a component of numerous fitness products. The 2'-deoxyinosine (**104**) is a purine 2'-deoxyribonucleoside. It is used in association with 5-fluorouracil, a chemotherapy medication, to enhance its activity. The 5-fluorouracil is an important therapy treatment for colorectal cancer, yet the overall response rate is less than 20%. In an attempt to increase the activity of 5-fluorouracil, it was found that 2'-deoxyinosine (**104**) has the ability to potentiate its cytotoxicity. Affiliation of 2'deoxyinosine (**104**) to 5-fluorouracil markedly increased the sensitivity of human colon carcinoma cell lines HT29 and SW620 by 38 and 700 times, respectively (Ciccolini et al., 2000). The 5-inosinic acid (**105**; Figure 22) is a nucleotide found in *Corynebacterium ammoniagenes*, *Synechococcus elongatus*, and other organisms. Although it is typically obtained from chicken by-products or other meat industry waste. 5-inosinic acid (**105**) plays an important role in metabolism. Furthermore, it is the first nucleotide formed during the purine nucleotide synthesis (Liu et al., 2012). The 5'-inosinic acid (**105**) is used as a flavour potentiator in a variety of foods due to its distinctive taste.

4.3. Pyrimidine Antibiotics

4.3.1. Nonglycosylated Pyrimidine Antibiotics

Bacimethrin (**106**) is the simplest antibiotic, isolated from *Bacillus megatherium* and from *Streptomyces albus* as illustrated in Figure 23 (Drautz et al., 1987). It is a naturally occurring thiamine antimetabolite. Bacimethrin (**106**) is effective *in vivo* and *in vitro* against *staphylococcal* infections as well as a number of yeasts and bacteria (Tanaka et al., 1962). Bacimethrin (**106**) has been shown to have some anticarcinoma activity in mice. It was also efficient against specific yeasts and bacteria in synthetic media, while it was ineffective or barely active in media having natural components like peptone, meat extract, yeast extract, etc. Significant antagonistic effects against bacimethrin (**106**) are demonstrated by vitamins B1 and B6.

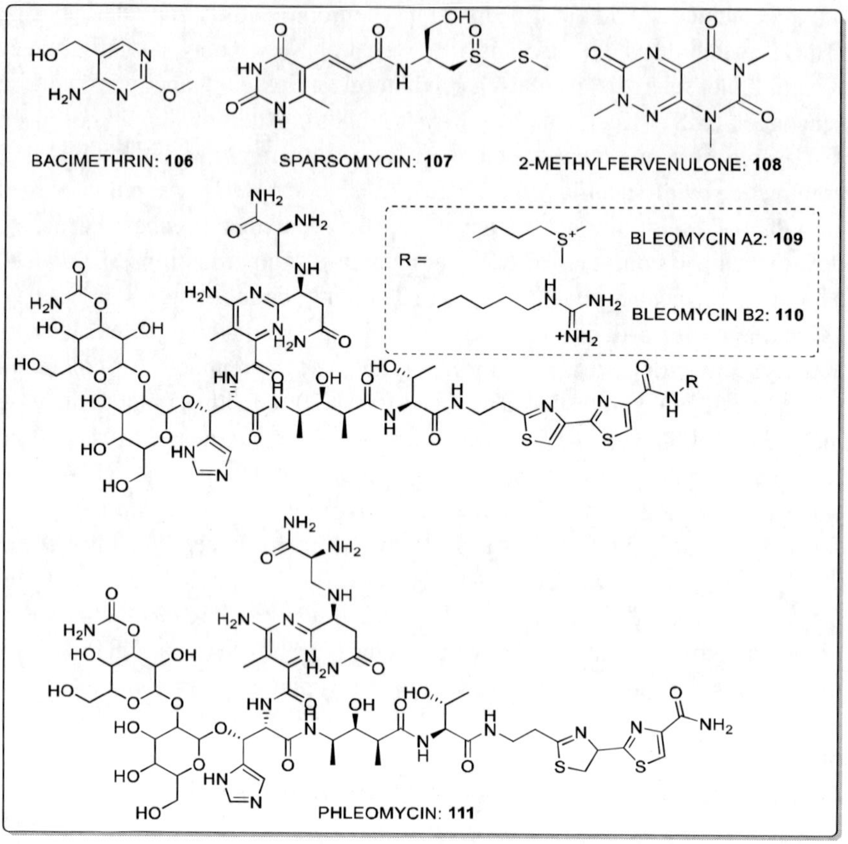

Figure 23. Nonglycosylated Pyrimidine Antibiotics.

Sparsomycin (**107**; Figure 23) is a non-glycosylated antibiotic and it is an inhibitor of the protein biosyntesis found in *Streptomyces cusidosporus* and *Streptomyces sparsogenes* (Porse et al., 1999). It is one of the few antibiotics that has the ability to inhibit peptidyl transferase, preventing bacteria from synthesising proteins. As a peptidyl transferase inhibitor, sparsomycin (**107**) prevents the formation of peptide bonds, which is a crucial step in protein biosynthesis that occurs on the large ribosomal subunit (Lazaro et al., 1991). The peptidyl transferase core is thought to contain a generally conserved structural motif that sparsomycin (**107**) detects. This motif may include pieces of ribosomal protein and/or rRNA. It may act by stabilising this interaction and thus restricting the movement of the 3'-end of the tRNA during elongation. Since, this motif is probably to be engaged in binding of the 3'-terminal adenosine of the P/P'-site I. Earlier, it was discovered that sparsomycin (**107**) could induce retinopathy, the possibility of it being an effective anti-tumor agent was initially considered.

2-Methylfervenulone (**108**; Figure 23) is a natural product that occurs in *Streptomyces* (Wang et al., 2000). It is a wide-spectrum antibiotic that also inhibits a number of protein tyrosine phosphatases (PTPs), a class of proteins frequently targeted in treating various diseases.

Bleomycins (**109-110**; Figure 23) and related glycopeptides are antibiotics obtained from the bacteria *Streptomyces verticillus* (Boger et al., 1996). Two major forms of bleomycins are differing in the positively charged tail ~60% bleomycin A2 (**109**) and ~30% bleomycin B2 (**110**). They have extraordinary structural properties that serve as the foundation for a novel method of antitumor action. Bleomycins (**109-110**) are frequently utilised in clinic because of its rapid attack on solid tumors. Furthermore, it is one of the few anticancer drugs that doesn't affect the bone marrow. Bleomycins (**109-110**) has the capability to disrupt single-stranded DNA in tumors and impede repair (Remers, 1984). Bleomycin's anticancer action is thought to be related to the presence of 9-(3-oxoprop-1-enyl)guanine. Bleomycins (**109-110**) are generally used to treat Hodgkin's lymphoma, non-Hodgkin's lymphoma, ovarian cancer, cervical cancer, and testicular cancer. Typically, it is injected intravenously into a muscle or peritoneal. To prevent cancer-related pleural effusion from recurring, bleomycins (**109-110**) can also be injected into the chest. Fever, vomiting, weight loss and rash are among Bleomycin's most frequent side effects. The most severe complication of bleomycins (**109-110**) is pulmonary fibrosis, which increases with increasing dosage and results in reduced lung function. Phleomycin (**111**), like bleomycin, is a glycopeptide antibiotic isolated from *Streptomyces verticillus* (Figure 23). It is a DNA synthesis inhibitor in bacteria (Maeda et al., 1956) and have been found to act as an antitumor agent (He et al., 1996). It binds and intercalates DNA, disrupting the double helix's integrity. Most bacteria, filamentous fungus, yeast, plant, and animal cells are susceptible to phleomycin (**111**). Phleomycin (**111**) prevents S-phase entry in the cell cycle. Despite the fact that phleomycin (**111**) can damage DNA in a manner similar to bleomycin (**110**), it is not employed as an anticancer drug. In experiments involving molecular genetics, phleomycin (**111**) is employed as a selective agent. Phleomycin (**111**) can be beneficial for identifying and selecting a number of cell types having a phleomycin (**111**) resistant gene, such as the Sh ble gene, due to its broad-spectrum toxicity against bacteria, plant, and animal cells.

4.3.2. Glycosylated Pyrimidine Antibiotics

Polyoxins (**112-123**) and neopolyoxins or nikkomycins (**124-129**) are a cluster of antibiotics produced by *Streptomycetes* which are used to treat fungal infection as depicted in Figure 24-26. Polyoxins (**112-123**) broadly can be structurally subcategorized into dipeptidyl polyoxins (**112-119**; Figure 24) and tripeptidyl polyoxins (**120-123**; Figure 25). They are the most well-known and efficient chitin synthase inhibitors (Zhang and Miller, 1999). As chitin synthase inhibitor, they have antifungal properties but no anti-bacterial properties. Nikkomycins (**124-129**) and polyoxins (**112-123**) are obtained from *Streptomyces tendae* and *Streptomyces cacoi* var. asoenis culture broths, respectively. Although highly harmful to insects and phytopathogenic fungi, vertebrates are not affected by polyoxins (**112-123**). Mammalian systems do not exhibit noticeable toxicity, which makes them appealing as possible agents against systemic fungi infections. Because of their failure to enter the cell and reach the chitin synthase site, polyoxins are ineffective against medically significant fungus like *Candida albicans* (Isono, 1988). It is corroborated by the finding that although the polyoxins (**112-123**) had similar effects on chitin synthetase preparations from various yeasts, including *Saccharomyces cerevisiae*, they only have an impact on *Saccharomyces cerevisiae* in culture when they are present in high concentrations. However, dipeptidyl polyoxins D (**113**) did affect *Saccharomyces cerevisiae* at lower concentrations when combined with medicines that permeabilize membranes (Endo and Misato, 1969). These findings with the apparent wide range of chitin-containing species that are vulnerable to polyoxin inhibition raised the possibility that may be active against *C. albicans* chitin synthase but are unable to penetrate the cell membrane to affect the enzyme there (Bowers et al., 1974). The dipeptidyl polyoxin L (**118**) has potent antifungal effect against *Candida sp.* (Shenbagamurthi et al., 1983).

The Nikkomycins (**124-129**; Figure 26), potent competitive inhibitors of chitin synthetases from fungi and yeasts, including *Candida albicans*, are closely linked to the polyoxins (**112-123**). In comparison to polyoxins (**112-123**), nikkomycins (**124-129**) are more effective against *C. albicans*, with nikkomycin Z (**127**) being the most effective natural analogue.

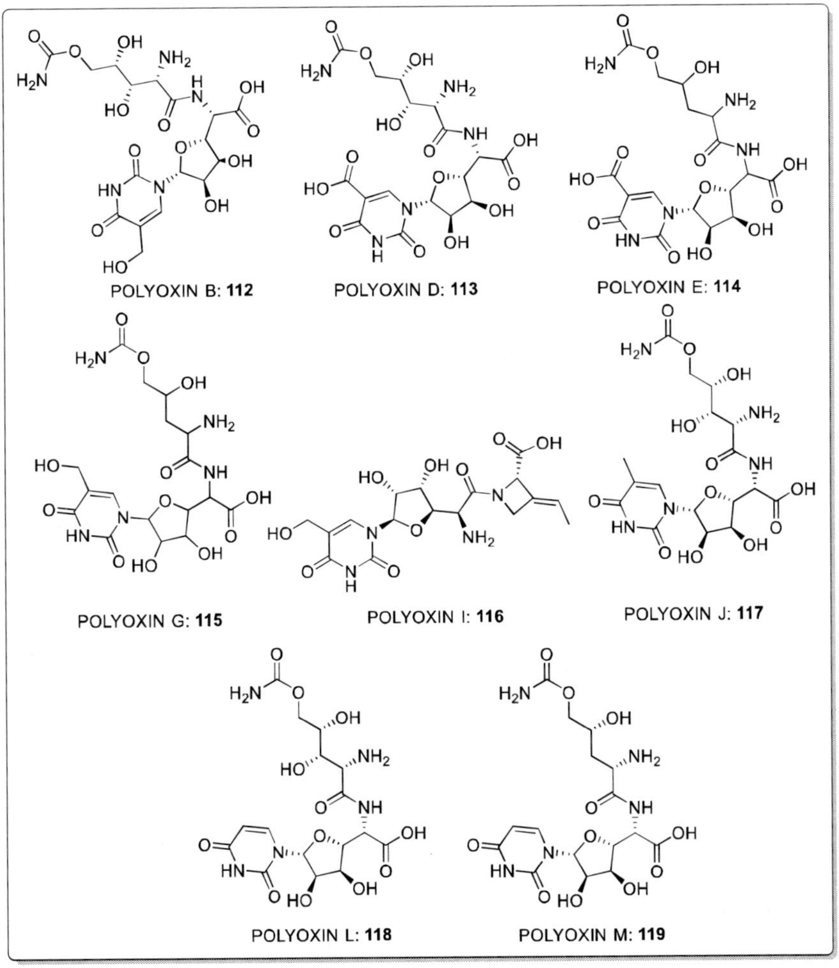

Figure 24. Glycosylated dipeptidyl pyrimidine Antibiotics.

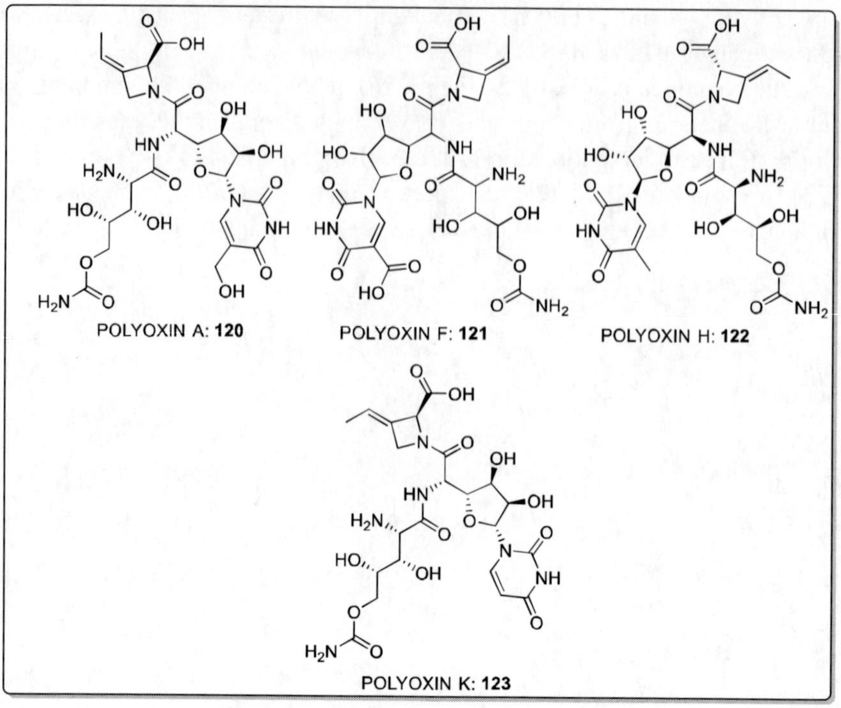

Figure 25. Glycosylated tripeptidyl pyrimidine Antibiotics.

Octosyl acid A (**130**) is yielded from *Streptomyces cacaoi* var. *asoenis* and it is an anhydrooctose as shown in Figure 27 (Danishefsky and Hungate, 1986). It shares structural similarities with ezomycin-class antifungals. The octosyl acid A (**130**) trans-glycosylation product, in which the 5-carboxyuracil is changed to an adenine base, is a potent inhibitor of cyclic-AMP phosphodiesterases from numerous animal tissues. Gougerotin (**131**) is a glycosylated aminoacyl-nucleoside antibacterial antibiotic isolated from *Streptomyces gougerotii* as depicted in Figure 27 (Fox et al., 1966; Kanzaki et al., 1962). By obstructing the transmission of amino acids from amino acyl tRNA to polypeptide, it suppresses protein synthesis (Clark Jr and Gunther, 1963). Additionally, it is a powerful inhibitor of the protein synthesis and elongation caused by auxin in *Avena coleoptiles*. Gougerotin (**131**) allows a difference to be made between these two stages of protein synthesis by selectively inhibiting the amalgamation of amino acids into protein while not affecting the release of completed protein chains from ribosomes. It was tested for its ability to disrupt the relative positioning of the 3'-terminal adenosine of

P/P'-site-bound tRNA and the peptidyl transferase loop region of 23S rRNA in the ribosome as an antibiotic inhibitor working within the peptidyl transferase center (Kirillov et al., 1999). The growth of certain viruses are also inhibited by gougerotin.

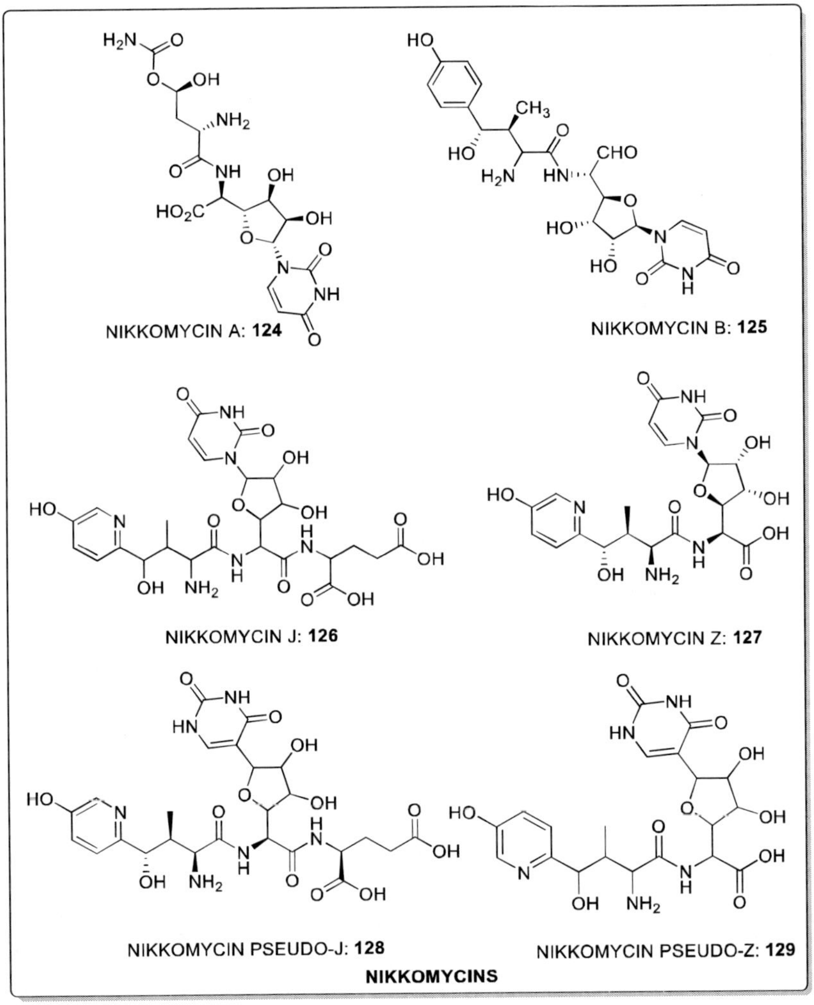

Figure 26. Nikkomycin alkaloids.

Bamicetin (**132**), amicetin (**133**), oxamicetin (**134**) and plicacetin (**135**) are related glycosylated antibiotics as shown in Figure 27. These antibiotics demonstrate cytosine as the pyrimidine base and an amino acid side chain. They function as inhibitors of peptidyl transferase B which is the larger subunit (Celma et al., 1970). These inhibitors are antagonized by Elongation factor G (EF-G) with GTP (Spirin and Asatryan, 1976). Furthermore, amicetin (**133**) is produced by *Streptomyces vinaceusdrappus* and *S. fascicularis*, and it protects HeLa cell monolayers infected with HSV-1 (Alarcón et al., 1984). Alternatively, oxamicetin (**134**) has been proved to be slightly more potent against gram-negative bacteria while being slightly less active against acid-fast and gram-positive bacteria (Lichtenthaler et al., 1975). Oxamicetin (**134**) is isolated from *Arthrobacter oxamicetus*. bamicetin (**132**) and plicacetin (**135**) are isolated from *Streptomyces plicatus* (Figure 27). Bamicetin (**132**) and plicacetin (**135**) are active against gram-positive bacteria and bamicetin (**132**) being more water soluble, exhibits the most activity against *E. coli*, whereas plicacetin (**135**) possesses less *in vivo* and *in vitro* activity than bamicetin (**132**).

Blasticidin S (**136**; Figure 27) is isolated from *Streptomyces griseochromogenes* as an antibiotic that has been widely utilized as a fungicide against *Pyricularia oryza*. Cytosinine and *L*-blastidic acid (Nomoto and Shimoyama, 2001) make up the structure of blasticidin S (**136**) (Kondo et al., 1973). In both prokaryotic and eukaryotic cells, Blasticidin S (**136**) prevents the biosynthesis of proteins (Yamaguchi et al., 1965). By interfering with protein translation, it works against bacteria, fungi and human cells. It is frequently utilized in cell culture for selecting and maintaining genetically manipulated cells.

Spongouridine (**137**) and spongothymidine (**138**) are arabinonucleosides that were initially discovered in the Caribbean sponge *Cryptotethia crypta* and later in the acorn worm *Ptychodera flava* as depicted in Figure 27 (Bergmann and Feeney, 1951). They are 1-β-arabinonucleosides of uracil and thymine, respectively. In cell culture, spongothymidine (**138**) prevents the replication of a few DNA viruses, including HSV-1, HSV-2, vaccinia (Kit et al., 1962), varicella-zoster (Dobersen et al., 1976), and pseudo arabies, but not RNA or cytomegalovirus (CMV) viruses (Renis and Buthala, 1965). Additionally, it prevents cell-free extracts made from cultures that are infected with HSV-1 from phosphorylating thymidine and deoxycytidine (Gentry and Aswell, 1975). Spongothymidine (**138**) is cytotoxic for L cells but not TK cells, while both host-cell DNA and viral DNA synthesis are impeded in HSV-1 infected cultures, DNA synthesis in infected cells is not noticeably impacted. It also prevents equine HSV-1 from growing in cell culture.

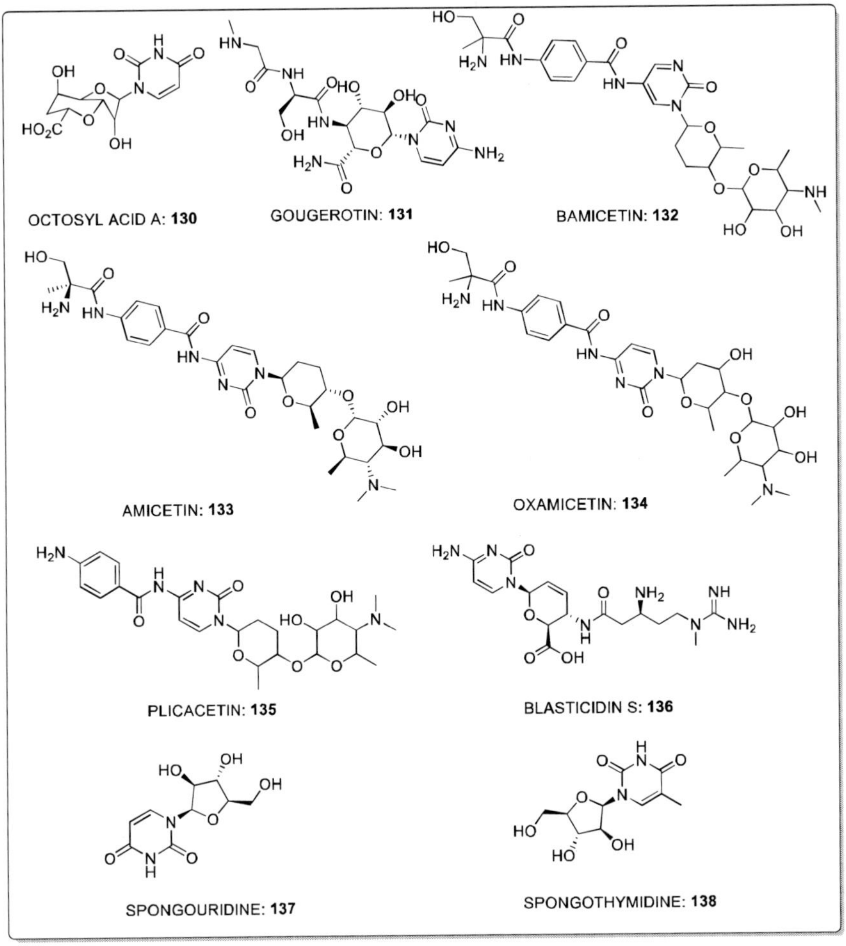

Figure 27. Miscellaneous glycosylated antibiotics.

5. Pyrimidine Vitamins

Folic acid (**139**), a water-soluble B vitamin (vitamin B9), can be found in fruits, peas, dried beans, leafy green vegetables, and nuts as well as cereals, enriched breads, and other grain products as mentioned in Figure 28 (Welch, 1983). It is used in the treatment of folic acid (**139**) deficiency and megaloblastic anaemia. Folic acid (**139**) not only prevents neural tube defects such as spina bifida, (Bibbins-Domingo et al., 2017) but it also can reduce homocysteine, which has a good effect on cardiovascular disease (Li et al., 2016; Tian et al., 2017). Beyond these clinical conditions, the role of this B vitamin includes resistance to birth defects, various cancers, Down's syndrome, affective disorders, dementia and serious conditions impairing pregnancy results. Prevention of birth abnormalities of the baby's brain and spine can be done by getting enough folic acid (**139**) before and during pregnancy. Additionally, it supports the body's ability to create new, healthy cells (Atta et al., 2016).

Riboflavin (**140**; Figure 28) is a water-soluble vitamin B (vitamin B2) that can be found in milk, eggs, cereals, mushrooms, and plain yogurt (Northrop-Clewes and Thurnham, 2012). Riboflavin (**140**) is also found in meat and fish, and many fruits and vegetables, particularly dark-green vegetables have significant levels of vitamin B2. Flavin mononucleotide and flavin adenine dinucleotide, two important coenzymes helped in cellular respiration, energy metabolism, antibody synthesis, and normal growth and development, are formed with the help of riboflavin (**140**). Riboflavin (**140**) is prescribed for migraines, prevents vitamin B2 deficiency, and is crucial for lowering excessive levels of homocysteine in the blood. A sufficient diet and supplementation with riboflavin appear to protect against sepsis, ischemia, and other conditions, while also lowering the likelihood of some cancers in humans (Suwannasom et al., 2020).

Thiamine (**141**) or vitamin B1 was the first vitamin B to have been recognized as shown in Figure 28 (Fattal-Valevski, 2011). It is a necessary nutrient that works with many energy-metabolizing enzymes as a cofactor. Sources of thiamine (**141**) include peas, fresh fruits (such as bananas and oranges), nuts, beef, poultry, cereals and beans. Thiamine (**141**) deficiency can result in Wernicke-Korsakoff syndrome and beriberi (Dhir et al., 2019). Thiamine (**141**) is a water-soluble vitamin that enters the bloodstream through the digestive system. It then travels through the blood and is eventually excreted through urination. Thiamine diphosphokinase enzyme transforms thiamine (**141**) in the blood to its active form, thiamine pyrophosphate (TPP). TPP is involved in various critical metabolic processes, including glycolysis, the Krebs cycle, and the pentose phosphate pathway.

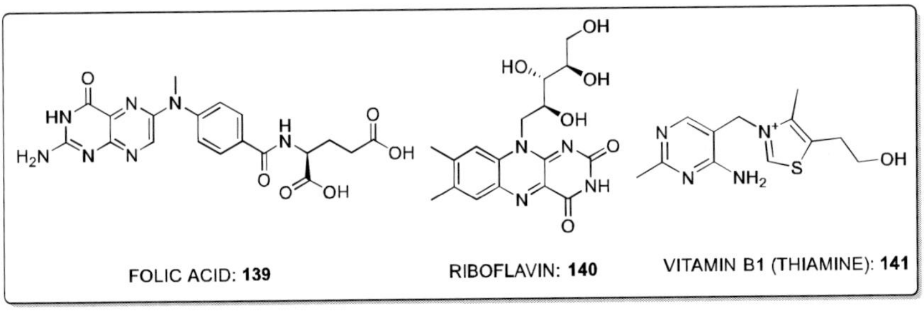

Figure 28. Pyrimidine ring containing vitamins.

6. Pyrimidine Toxins

Cylindrospermopsin (**142**) is a hepatotoxin produced by a freshwater alga *Cylindrospermopsis Raciborskii* as shown in Figure 29 (Armah et al., 2013). The occurrence of this compound in freshwater raises concerns not only for human health but also for the well-being of the entire ecosystem. Tropical Australia has had a serious hepatoenteritis outbreak as a result of cylindrospermopsin (**142**) production by algal blooms in drinking water. Early symptoms of cylindrospermopsin (**142**) toxicity include anorexia, hepatomegaly, constipation, headache, vomiting, fever, and abdominal pain. Hematuria was a common symptom among patients, and in about 80% of kids who had cylindrospermopsin (**142**) toxicity, the symptoms worsened to include hypokalemia, acidosis, and severe electrolyte abnormalities. Symptoms of toxicity in the hepatic and renal systems have been observed in the later phases of toxicity. The affected children suffered hyperemic or bleeding mucous membranes, acidotic shock, and bloody diarrhea that continued for three weeks in some instances. Due to these consequences, cylindrospermopsin (**142**) has been demonstrated to be hepatotoxic, cytotoxic, toxic to foetuses, genotoxic and likely carcinogenic. While isolating cylindrospermopsin (**142**) for biological activities, a minor component of the hydrophilic extract known as 7-epicylindrospermopsin (**143**; Figure 29), (Banker et al., 2000) which is produced by the cyanobacterium *Aphanizomenon ovalisporum*. 7-Epicylindrospermopsin (**143**) is equally toxic as cylindrospermopsin (**142**) (White and Hansen, 2005). Closely related to epicylindrospermopsins (**143**) are ptilocaulins (**144**) and isoptilocaulins (**145**). Ptilocaulins (**144**) are extracted from *Ptilocaulis aff* (Harbour et al., 1981), *Ptilocaulis spiculifer*, a Caribbean sponge. They have high antibacterial activity against both gram-positive and gram-negative bacteria, filamentous fungi and yeasts. Along with cylindrospermopsin (**142**) and 7-epicylindrospermopsin (**143**), ptilocaulins (**144**) and isoptilocaulins (**145**) are hepatotoxins (Figure 29).

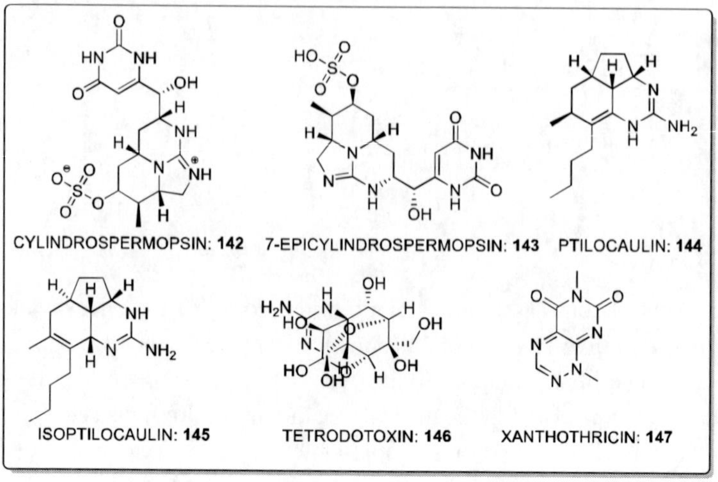

Figure 29. Pyrimidine ring containing toxins.

Tetrodotoxin (TTX) (**146**) is one of the most potent non-protein neurotoxins as depicted in Figure 29 (Chau et al., 2011). It is present in the ovaries and livers of the Japanese Puffer fish, as well as the *California Newt* or Salamander. Notably, it is now recognized that TTX (**146**) is yielded by symbiotic bacteria such as *Pseudoalteromonas*, *Pseudomonas*, and others. TTX (**146**) is a sodium channel blocker that prevents the firing of action potentials in neurons, causing rapid weakening and paralysis of muscles, including those of the respiratory tract, resulting in respiratory arrest and death (Lago et al., 2015). The early signs of TTX (**146**) poisoning are headache, numbness, excessive sweating (diaphoresis), salivation (ptyalism), dizziness, nausea, diarrhoea, vomiting (emesis), abdominal (epigastric) discomfort, difficulty moving (motor dysfunction), weakness (malaise), and trouble speaking. The second stage of the poisoning is characterized by progressive paralysis of the entire body, culminating in the respiratory muscles which results in death.

Toxoflavin (xanthothricin) (**147**) is a toxin produced by several bacteria, including *Burkholderia gladioi* and *Streptomyces himensis* (Figure 29). It also has antibacterial capabilities and can function as a pH indicator. It functions as an antineoplastic agent, an apoptosis inducer, a bacterial metabolite, an antibacterial agent, and a virulence factor (Machlowitz, 1954).

Conclusion

The pyrimidine alkaloids discussed in this book chapter have been shown to possess important biological activities such as antifungal, antibacterial, anticancer, ant-aging, antiviral, antitumor, antiulcer, antidepressant and antiparasitic etc. Furthermore, the analogs of some of the pyrimidine ring-containing alkaloids have also been discussed that have exhibited important activities or exceptional medicinal qualities. On account of their biological activity, some alkaloids have been marketed as drugs or some of them is used as lead for new drug discovery process. The pyrimidine ring-containing alkaloids have been demonstrated as a major contributor to drug research in this chapter. Their importance in this field can only be expected to rise in the future.

Future Prospectives

Alkaloids, despite having a well-established role in drug discovery, there is still a lot to learn about them. The potential for the identification of novel alkaloids in the terrestrial and marine biomes is tremendous and largely untapped. These alkaloids will be necessary to address worldwide disease requirements. To satisfy the needs for the treatment of rare and neglected tropical diseases, multidrug-resistant diseases, tuberculosis, and malaria, a new universal paradigm for natural product discovery and development is essential. New, repurposed, and modified alkaloids will be extremely helpful for these and other medication development endeavors. Recent advancements in computational (Artificial intelligence and machine learning) and hyphenated-analytical methods have created new prospects for identifying the complex natural products structure to create novel medications. A brand-new and interesting era in alkaloid research has already begun.

References

Adibhatla RM, Hatcher J, Dempsey R. Citicoline: neuroprotective mechanisms in cerebral ischemia. *Journal of Neurochemistry* (2002) 80:12-23.

Agteresch HJ, Dagnelie PC, van den Berg JWO, Wilson J. Adenosine triphosphate. *Drugs* (1999) 58:211-232.

Ahmad, Farediah, Hazar B. Mohd Ismail and Mawardi Rahmani, (1996). Arborine, a Larval Growth Inhibitor from Glycosmis pentaphylla. *Pertanika J. Sci. & Technol.* 4(1): 11-15.

Aksoy SÇ, Uzel A, Bedir E. Cytosine-type nucleosides from marine-derived Streptomyces rochei 06CM016. *The Journal of Antibiotics* (2016) 69:51-56.

Alarcón B, Lacal JC, Fernández-Sousa J, Carrasco L. Screening for new compounds with antiherpes activity. *Antiviral Research* (1984) 4:231-244.

Armah de la Cruz A, Anastasia Hiskia, Triantafyllos Kaloudis, Neil Chernoff, Donna Hill, Maria G Antoniou, Xuexiang He, Keith Loftin, Kevin O'Shea, Cen Zhao, Miguel Pelaez, Changseok Han, Trevor J Lynch and Dionysios D Dionysiou. A review on cylindrospermopsin: the global occurrence, detection, toxicity and degradation of a potent cyanotoxin. *Environmental Science: Processes & Impacts* (2013) 15:1979-2003.

Asami, Yukihiro, Jae-Hyuk Jang, Nak-Kyun Soung, Long He, Dong Oh Moon, Jong Won Kim, Hyuncheol Oh, Makoto Muroi, Hiroyuki Osada, Bo Yeon Kim, Jong Seog Ahn. Protuboxepin A, a marine fungal metabolite, inducing metaphase arrest and chromosomal misalignment in tumor cells. *Bioorganic & Medicinal Chemistry* (2012) 20:3799-3806.

Atta, Callie AM, Kirsten M Fiest, Alexandra D Frolkis, Nathalie Jette, Tamara Pringsheim, Christine St Germaine-Smith, Thilinie Rajapakse, Gilaad G Kaplan, Amy Metcalfe. Global birth prevalence of spina bifida by folic acid fortification status: a systematic review and meta-analysis. *American Journal of Public Health* (2016) 106:e24-e34.

Banker R, Teltsch B, Sukenik A, Carmeli S. 7-Epicylindrospermopsin, a Toxic Minor Metabolite of the Cyanobacterium Aphanizomenon o valisporum from Lake Kinneret, Israel. *Journal of Natural Products* (2000) 63:387-389.

Bergmann W, Feeney RJ. Contributions to the study of marine products. XXXII. The nucleosides of sponges. I. *The Journal of Organic Chemistry* (1951) 16:981-987.

Bibbins-Domingo, Kirsten, David C Grossman, Susan J Curry, Karina W Davidson, John W Epling Jr, Francisco AR García, Alex R Kemper, Alex H Krist, Ann E Kurth, C Seth Landefeld, Carol M Mangione, William R Phillips, Maureen G Phipps, Michael P Pignone, Michael Silverstein, Chien-Wen Tseng, US Preventive Services Task Force. Folic acid supplementation for the prevention of neural tube defects: US Preventive Services Task Force recommendation statement. *JAMA* (2017) 317:183-189.

Blackburn, G Michael, Martin Egli, Michael J Gait, Jonathan K Watts. *Nucleic Acids in Chemistry and Biology.* (2006): Royal Society of Chemistry.

Boger DL, Teramoto S, Cai H. Synthesis and evaluation of deglycobleomycin A2 analogues containing a tertiary N-methyl amide and simple ester replacement for the l-histidine secondary amide: Direct functional characterization of the requirement for secondary amide metal complexation. *Bioorganic & Medicinal Chemistry* (1996) 4:179-193.

Borthwick AD. 2, 5-Diketopiperazines: synthesis, reactions, medicinal chemistry, and bioactive natural products. *Chemical Reviews* (2012) 112:3641-3716.

Bos JL. Epac proteins: multi-purpose cAMP targets. *Trends in Biochemical Sciences* (2006) 31:680-686.

Bowers B, Levin G, Cabib E. Effect of polyoxin D on chitin synthesis and septum formation in Saccharomyces cerevisiae. *Journal of Bacteriology* (1974) 119:564-575.

Buchanan BB, Gruissem W, Jones RL. Biochemistry and Molecular Biology of Plants. (2015): John wiley & sons.

Celma M, Monro R, Vazquez D. Substrate and antibiotic binding sites at the peptidyl transferase centre of E. coli ribosomes. *FEBS Letters* (1970) 6:273-277.

Chau R, Kalaitzis JA, Neilan BA. On the origins and biosynthesis of tetrodotoxin. *Aquatic Toxicology* (2011) 104:61-72.

Chen P, Goldberg DE, Kolb B, Lanser M, Benowitz LI. Inosine induces axonal rewiring and improves behavioral outcome after stroke. *Proceedings of the National Academy of Sciences* (2002) 99:9031-9036.

Choudhary A, Naughton LM, Montánchez I, Dobson AD, Rai DK. Current status and future prospects of marine natural products (MNPs) as antimicrobials. *Marine Drugs* (2017) 15:272.

Ciccolini, Joseph, Laurent Peillard, Alexandre Evrard, Pierre Cuq, Claude Aubert, André Pelegrin, Patricia Formento, Gérard Milano, and Jacques Catalin. Enhanced antitumor activity of 5-fluorouracil in combination with 2′-deoxyinosine in human colorectal cell lines and human colon tumor xenografts. *Clinical Cancer Research* (2000) 6:1529-1535.

Clark Jr J, Gunther JK. Gougerotin, a specific inhibitor of protein synthesis. *Biochimica et Biophysica Acta (BBA)-Specialized Section on Nucleic Acids and Related Subjects* (1963) 76:636-638.

Costa, Emmanoel V, Maria Lúcia B Pinheiro, Clahildek M Xavier, Jefferson RA Silva, Ana Cláudia F Amaral, Afonso DL Souza, Andersson Barison, Francinete R Campos, Antonio G Ferreira, Gérzia MC Machado and Leonor LP Leon. A Pyrimidine-β-carboline and Other Alkaloids from Annona foetida with Antileishmanial Activity. *Journal of Natural Products* (2006) 69:292-294.

Costa, Renyer A, Earle Silva A Junior, Guilherme Braule P Lopes, Maria Lucia B. Pinheiro, Emmanoel V Costa, Daniel P Bezerra, Kelson Oliveira. Structural, vibrational, UV–vis, quantum-chemical properties, molecular docking and anti-cancer activity study of annomontine and N-hydroxyannomontine β-carboline alkaloids: a combined experimental and DFT approach. *Journal of Molecular Structure* (2018) 1171:682-695.

Danishefsky S, Hungate R. The total synthesis of octosyl acid A: a new departure in organostannylene chemistry. *Journal of the American Chemical Society* (1986) 108:2486-2487.

Dhir S, Tarasenko M, Napoli E, Giulivi C. Neurological, psychiatric, and biochemical aspects of thiamine deficiency in children and adults. *Frontiers in Psychiatry* (2019):207.

Dobersen MJ, Jerkofsky M, Greer S. Enzymatic basis for the selective inhibition of varicella-zoster virus by 5-halogenated analogues of deoxycytidine. *Journal of Virology* (1976) 20:478-486.

Dobolyi A, Juhász G, Kovács Z, Kardos J. Uridine function in the central nervous system. *Current Topics in Medicinal Chemistry* (2011) 11:1058-1067.

Dong, Ji, Shi-sheng Huang, Ya-nan Hao, Zi-wen Wang, Yu-xiu Liu, Yong-qiang Li, Qing-min Wang. Marine-natural-products for biocides development: first discovery of meridianin alkaloids as antiviral and anti-phytopathogenic-fungus agents. *Pest Management Science* (2020) 76:3369-3376.

Doveston RG, Steendam R, Jones S, Taylor RJ. Total synthesis of an oxepine natural product,(±)-janoxepin. *Organic Letters* (2012) 14:1122-1125.

Drautz H, Messerer W, Zähner H, Breiding-Mack S, Zeeck A. Metabolic products of microorganisms. 239 bacimethrin isolated from Streptomyces albus identification, derivatives, synthesis and biological properties. *The Journal of Antibiotics* (1987) 40:1431-1439.

Dreyer DL, Brenner R. Alkaloids of some Mexican Zanthoxylum species. *Phytochemistry* (1980) 19:935-939.

El-Gengaihi S, Karawya M, Selim M, Motawe H, Ibrahim N, Faddah L. A novel pyrimidine glycoside from Momordica charantia L. *Pharmazie* (1995) 50:361-362.

Endo A, Misato T. Polyoxin D, a competitive inhibitor of UDP-N-acetylglucosamine: chitin N-acetylglucosaminyltransferase in Neurospora crassa. *Biochemical and Biophysical Research Communications* (1969) 37:718-722.

Fattal-Valevski A. Thiamine (vitamin B1). *Journal of Evidence-Based Complementary & Alternative Medicine* (2011) 16:12-20.

Fox JJ, Watanabe KA, Bloch A. Nucleoside antibiotics. *Progress in Nucleic Acid Research and Molecular Biology* (1966) 5:251-313.

Gentry GA, Aswell JF. Inhibition of herpes simplex virus replication by araT. *Virology* (1975) 65:294-296.

Gompel, Marie, Maryse Leost, Elisa Bal De Kier Joffe, Lydia Puricelli, Laura Hernandez Franco, Jorge Palermo, Laurent Meijer. Meridianins, a new family of protein kinase inhibitors isolated from the ascidian Aplidium meridianum. *Bioorganic & Medicinal Chemistry Letters* (2004) 14:1703-1707.

Goyal A, Senst B, Bhyan P, Zeltser R. Reentry Arrhythmia. (2019).

Gullo, V, M Conover, R Cooper, C Federbush, AC Horan, T Kung, J Marquez, M Patel, A Watnick. SCH 36605, a novel anti-inflammatory compound taxonomy, fermentation, isolation and biological properties. *The Journal of Antibiotics* (1988) 41:20-24.

Guo JU, Su Y, Zhong C, Ming G-l, Song H. Hydroxylation of 5-methylcytosine by TET1 promotes active DNA demethylation in the adult brain. *Cell* (2011) 145:423-434.

Gupta O, Anand K, Ghatak B, Atal C. Vasicine, alkaloid of Adhatoda vasica, a promising uterotonic abortifacient. *Indian Journal of Experimental Biology* (1978).

Gupta O, Sharma M, Ghatak B, Atal C. Potent uterine activity of alkaloid vasicine. *The Indian Journal of Medical Research* (1977) 66:865-871.

Harbour, Gary C, Adrienne A Tymiak, Kenneth L Rinehart Jr, Paul D Shaw, Robert Hughes Jr, Stephen A Mizsak, John H Coats, Gary E Zurenko, Li H Li, and Sandra L Kuentzel. Ptilocaulin and isoptilocaulin, antimicrobial and cytotoxic cyclic guanidines from the

Caribbean sponge Ptilocaulis aff. P. spiculifer (Lamarck, 1814). *Journal of the American Chemical Society* (1981) 103:5604-5606.

Hassett A, Blaettler W, Knowles JR. Pyruvate kinase: is the mechanism of phospho transfer associative or dissociative? *Biochemistry* (1982) 21:6335-6340.

Hayashida M, Fukuda K-i, Fukunaga A. Clinical application of adenosine and ATP for pain control. *Journal of Anesthesia* (2005) 19:225-235.

He CH, Masson J-Y, Ramotar D. A Saccharomyces cerevisiae phleomycin-sensitive mutant, phl40, is defective in the RAD6 DNA repair gene. *Canadian Journal of Microbiology* (1996) 42:1263-1266.

Ignacimuthu S, Shanmugam N. Antimycobacterial activity of two natural alkaloids, vasicine acetate and 2-acetyl benzylamine, isolated from Indian shrub Adhatoda vasica Ness. leaves. *Journal of Biosciences* (2010) 35:565-570.

Isono K. Nucleoside antibiotics: structure, biological activity, and biosynthesis. *The Journal of Antibiotics* (1988) 41:1711-1739.

Jahn T, König GM, Wright AD, Wörheide G, Reitner J. Manzacidin D: An unprecedented secondary metabolite from the "living fossil" sponge Astrosclera willeyana. *Tetrahedron Letters* (1997) 38:3883-3884.

Jane DE, Hoo K, Kamboj R, Deverill M, Bleakman D, Mandelzys A. Synthesis of willardiine and 6-azawillardiine analogs: pharmacological characterization on cloned homomeric human AMPA and kainate receptor subtypes. *Journal of Medicinal Chemistry* (1997) 40:3645-3650.

Jiang, Suping, Qiang Zeng, Montip Gettayacamin, Anchalee Tungtaeng, Srisombat Wannaying, Apasorn Lim, Pranee Hansukjariya, Christopher O Okunji, Shuren Zhu, Daohe Fang. Antimalarial activities and therapeutic properties of febrifugine analogs. *Antimicrobial Agents and Chemotherapy* (2005) 49:1169-1176.

Johnson TB, Coghill RD. Researches on pyrimidines. C111. The discovery of 5-methylcytosine in tuberculinic acid, the nucleic acid of the tubercle bacillus1. *Journal of the American Chemical Society* (1925) 47:2838-2844.

Jonas, DA, I Elmadfa, KH Engel, KJ Heller, G Kozianowski, A König, D Müller, JF Narbonne, W Wackernagel, J Kleiner. Safety Considerations of DNA in Food. *Annals of Nutrition and Metabolism* (2001) 45:235-254.

Kafer, Chris, Lan Zhou, Djoko Santoso, Adel Guirgis, Brock Weers, Sanggyu Park, Robert Thornburg. Regulation of pyrimidine metabolism in plants. *Frontiers in Bioscience-Landmark* (2004) 9:1611-1625.

Kanzaki, Toshihiko, Eiji Higashide, Hiroichi Yamamoto, Motoö Shibata, Kôiti Nakazawa, Hidesuke Iwasaki, Torao Takewaka, Akira Miyake. Gougerotin, a new antibacterial antibiotic. *The Journal of Antibiotics, Series A* (1962) 15:93-97.

Kim K-W, Roh JK, Wee H-J, Kim C. Molecular Targeted Anticancer Drugs. In: *Cancer Drug Discovery* (2016): Springer. 175-238.

Kirillov SV, Porse BT, Garrett RA. Peptidyl transferase antibiotics perturb the relative positioning of the 3'-terminal adenosine of P/P'-site-bound tRNA and 23S rRNA in the ribosome. *RNA* (1999) 5:1003-1013.

Kit S, Dubbs D, Piekarski LJ. Enhanced thymidine phosphorylating activity of mouse fibroblasts (strain LM) following vaccinia infection. *Biochemical and Biophysical Research Communications* (1962) 8:72-75.

Kobayashi, Jun'ichi, Jie-fei Cheng, Yumiko Kikuchi, Masami Ishibashi, Shosuke Yamamura, Yasushi Ohizumi, Tomihisa Ohtac, Shigeo Nozoec. Rigidin, a novel alkaloid with calmodulin antagonistic activity from the okinawan marine tunicate Eudistoma cf. rigida. *Tetrahedron Letters* (1990) 31:4617-4620.

Koch M, De Keyser J. Uric acid in multiple sclerosis. *Neurological Research* (2006) 28:316-319.

Koepfli J, Mead J, Brockman JA. Alkaloids of Dichroa febrifuga. I. Isolation and degradative studies. *Journal of the American Chemical Society* (1949) 71:1048-1054.

Kondo T, Nakai H, Goto T. Synthesis of cytosinine, the nucleoside component of antibiotic blasticidin S. *Tetrahedron* (1973) 29:1801-1806.

Kriaucionis S, Heintz N. The nuclear DNA base 5-hydroxymethylcytosine is present in Purkinje neurons and the brain. *Science* (2009) 324:929-930.

Lago J, Rodríguez LP, Blanco L, Vieites JM, Cabado AG. Tetrodotoxin, an extremely potent marine neurotoxin: Distribution, toxicity, origin and therapeutical uses. *Marine Drugs* (2015) 13:6384-6406.

Lappas CM. The plant hormone zeatin riboside inhibits T lymphocyte activity via adenosine A2A receptor activation. *Cellular & Molecular Immunology* (2015) 12:107-112.

Lazaro E, San Felix A, Van den Broek L, Ottenheijm H, Ballesta J. Interaction of the antibiotic sparsomycin with the ribosome. *Antimicrobial Agents and Chemotherapy* (1991) 35:10-13.

Leboeuf M, Cavé A, Forgacs P, Provost J, Chiaroni A, Riche C. Alkaloids of the Annonaceae. Part 33. Annomontine and methoxyannomontine, two new pyrimidine-β-carboline-type alkaloids from Annona montana. *Journal of the Chemical Society, Perkin Transactions 1* (1982):1205-1208.

Lee SU, Asami Y, Lee D, Jang J-H, Ahn JS, Oh H. Protuboxepins A and B and protubonines A and B from the marine-derived fungus Aspergillus sp. SF-5044. *Journal of Natural Products* (2011) 74:1284-1287.

Lewis RA. Hawley's condensed chemical dictionary. (2016): John Wiley & Sons.

Li Y, Huang T, Zheng Y, Muka T, Troup J, Hu FB. Folic acid supplementation and the risk of cardiovascular diseases: a meta-analysis of randomized controlled trials. *Journal of the American Heart Association* (2016) 5:e003768.

Liang X, Zhang X, Lu X, Zheng Z, Ma X, Qi S. Diketopiperazine-type alkaloids from a deep-sea-derived Aspergillus puniceus fungus and their effects on liver X receptor α. *Journal of Natural Products* (2019) 82:1558-1564.

Lichtenthaler F, Černá J, Rychlik I. The effect of oxamicetin and some amicetin analogs on ribosomal peptidyl transferase. *FEBS Letters* (1975) 53:184-187.

Lin Y-L, Huang R-L, Chang C-M, Kuo Y-H. Two new puriniums and three new pyrimidines from Heterostemma brownii. *Journal of Natural Products* (1997) 60:982-985.

Liu, F, S-W You, L-P Yao, H-L Liu, X-Y Jiao, M Shi, Q-B Zhao and G Ju. Secondary degeneration reduced by inosine after spinal cord injury in rats. *Spinal Cord* (2006) 44:421-426.

Liu Z-Q, Zhang L, Sun L-H, Li X-J, Wan N-W, Zheng Y-G. Enzymatic production of 5'-inosinic acid by a newly synthesised acid phosphatase/phosphotransferase. *Food Chemistry* (2012) 134:948-956.

Lu, Xin-Hua, Qing-Wen Shi, Zhi-Hui Zheng, Ai-Bing Ke, Hua Zhang, Chang-Hong Huo, Ying Ma, Xiao Ren, Ye-Ying Li, Jie Lin, Qin Jiang, Yu-Cheng Gu, Hiromasa Kiyota. Oxepinamides: Novel liver X receptor agonists from Aspergillus puniceus (2011): Wiley Online Library.

Luzzatto L, Arese P. Favism and glucose-6-phosphate dehydrogenase deficiency. *New England Journal of Medicine* (2018) 378:60-71.

Machado-Vieira, Rodrigo, Giacomo Salvadore, Nancy DiazGranados, Lobna Ibrahim, David Latov, Cristina Wheeler-Castillo, Jacqueline Baumann, Ioline D Henter, and Carlos A Zarate, Jr. New therapeutic targets for mood disorders. *The Scientific World Journal* (2010) 10:713-726.

Machlowitz R. Xanthothricin, a new antibiotic. *Antibiot Chemother* (1954) 4:259-261.

Maeda K, Kosaka H, Yagishita, H K. Umezawa: A new antibiotic, phleomycin. *The Journal of Antibiotics, Series A* (1956) 9:82-85.

Mitchell J, Lazarenko G. Wide QRS complex tachycardia. *Canadian Journal of Emergency Medicine* (2008) 10:572-578.

Mok DW, Mok MC. Cytokininschemistry, activity, and function. (1994): CRC press.

Moloudizargari M, Mikaili P, Aghajanshakeri S, Asghari MH, Shayegh J. Pharmacological and therapeutic effects of Peganum harmala and its main alkaloids. *Pharmacognosy Reviews* (2013) 7:199.

Murugan N, Srinivasan R, Murugan A, Kim M, Natarajan D. Glycosmis pentaphylla (Rutaceae): A natural candidate for the isolation of potential bioactive arborine and skimmianine compounds for controlling multidrug-resistant Staphylococcus aureus. *Frontiers in Public Health* (2020) 8:176.

Nishioka, H, M Imoto, T Sawa, M Hamada, H Naganawa, T Takeuchi, K Umezawa. Screening of phosphatidyl-inositol kinase inhibitors from Streptomyces. *The Journal of Antibiotics* (1989) 42:823-825.

Niwa H, Yoshida Y, Yamada K. A brominated quinazolinedione from the marine tunicate Pyura sacciformis. *Journal of Natural Products* (1988) 51:343-344.

Nomoto S, Shimoyama A. First synthesis of blastidic acid, a component amino acid in an antibiotic, blasticidin S. *Tetrahedron Letters* (2001) 42:1753-1755.

Northrop-Clewes CA, Thurnham DI. The discovery and characterization of riboflavin. *Annals of Nutrition and Metabolism* (2012) 61:224-230.

O'Keefe Jr JH, Bateman TM, Silvestri R, Barnhart C. Safety and diagnostic accuracy of adenosine thallium-201 scintigraphy in patients unable to exercise and those with left bundle branch block. *American Heart Journal* (1992) 124:614-621.

Oura, Hajimu, Masato Iino, Yosuke Nakazawa, Masahiro Tajima, Ritsuro Ideta, Yutaka Nakaya, Seiji Arase, Jiro Kishimoto. Adenosine increases anagen hair growth and thick hairs in Japanese women with female pattern hair loss: a pilot, double-blind, randomized, placebo-controlled trial. *The Journal of Dermatology* (2008) 35:763-767.

Pakrashi S, Bhattacharyya J. Studies on indian medicinal plants—XIV: Interrelationships among the quinazoline alkaloids from glycosmis arborea (Roxb.) DC. *Tetrahedron* (1968) 24:1-5.

Pakrashi S, Bhattacharyya J, Johnson L, Budzikiewicz H. Studies on indian medicinal plants—VI: Structures of glycosmicine, glycorine and glycosminine, the minor alkaloids from Glycosmis arborea (roxb.) DC. *Tetrahedron* (1963) 19:1011-1026.

Pascal JM. DNA and RNA ligases: structural variations and shared mechanisms. *Current Opinion in Structural Biology* (2008) 18:96-105.

Pauletti, Patrícia Mendonça, Lucas Silva Cintra, Caio Guedes Braguine, Ademar Alves da Silva Filho, Márcio Luís Andrade e Silva, Wilson Roberto Cunha, and Ana Helena Januário. Halogenated indole alkaloids from marine invertebrates. *Marine Drugs* (2010) 8:1526-1549.

Perry, Nigel B, Laurent Ettouati, Marc Litaudon, John W Blunt, Murray HG Munro, Sean Parkin, Hakon Hope. Alkaloids from the antarctic sponge Kirkpatrickia varialosa.: Part 1: Variolin b, a new antitumour and antiviral compound. *Tetrahedron* (1994) 50:3987-3992.

Porse BT, Kirillov SV, Awayez MJ, Ottenheijm HC, Garrett RA. Direct crosslinking of the antitumor antibiotic sparsomycin, and its derivatives, to A2602 in the peptidyl transferase center of 23S-like rRNA within ribosome-tRNA complexes. *Proceedings of the National Academy of Sciences* (1999) 96:9003-9008.

Ralifo P, Tenney K, Valeriote FA, Crews P. A distinctive structural twist in the aminoimidazole alkaloids from a calcareous marine sponge: isolation and characterization of leucosolenamines A and B. *Journal of Natural Products* (2007) 70:33-38.

Raman A, Lau C. Anti-diabetic properties and phytochemistry of Momordica charantia L.(Cucurbitaceae). *Phytomedicine* (1996) 2:349-362.

Rand TG, Giles S, Flemming J, Miller JD, Puniani E. Inflammatory and cytotoxic responses in mouse lungs exposed to purified toxins from building isolated Penicillium brevicompactum Dierckx and P. chrysogenum Thom. *Toxicological Sciences* (2005) 87:213-222.

Rattan SI, Sodagam L. Gerontomodulatory and youth-preserving effects of zeatin on human skin fibroblasts undergoing aging in vitro. *Rejuvenation Research* (2005) 8:46-57.

Remers WA. Antineoplastic agents. (1984): Krieger Publishing Company.

Renis H, Buthala D. Development of resistance to antiviral drugs. *Annals of the New York Academy of Sciences* (1965) 130:343-354.

Rizzello, Carlo Giuseppe, Ilario Losito, Laura Facchini, Kati Katina, Francesco Palmisano, Marco Gobbetti and Rossana Coda. Degradation of vicine, convicine and their aglycones during fermentation of faba bean flour. *Scientific Reports* (2016) 6:1-11.

Roggen H, Charnock C, Gundersen L-L. The first total synthesis of heteromine B, and an improved synthesis of heteromine A; natural products with antitumor activities. *Tetrahedron* (2009) 65:5199-5203.

Roué M, Quévrain E, Domart-Coulon I, Bourguet-Kondracki M-L. Assessing calcareous sponges and their associated bacteria for the discovery of new bioactive natural products. *Natural Product Reports* (2012) 29:739-751.

Scheeff ED, Bourne PE. Structural evolution of the protein kinase–like superfamily. *PLoS Computational Biology* (2005) 1:e49.

Shenbagamurthi P, Smith HA, Becker JM, Steinfeld A, Naider F. Design of anticandidal agents: synthesis and biological properties of analogs of polyoxin L. *Journal of Medicinal Chemistry* (1983) 26:1518-1522.

Simone, Matteo, Eugenio Erba, Giovanna Damia, Faina Vikhanskaya, Angela M Di Francesco, Riccardo Riccardi, Christian Bailly, Carmen Cuevas, José Maria Fernandez Sousa-Faro, Maurizio D'Incalci. Variolin B and its derivate deoxy-variolin B: new marine natural compounds with cyclin-dependent kinase inhibitor activity. *European Journal of Cancer* (2005) 41:2366-2377.

Soni S, Anandjiwala S, Patel G, Rajani M. Validation of different methods of preparation of Adhatoda vasica leaf juice by quantification of total alkaloids and vasicine. *Indian Journal of Pharmaceutical Sciences* (2008) 70:36.

Spirin A, Asatryan L. The effect of ribosomal peptidyl-transferase inhibitors is antagonized by elongation factor G with GTP. *FEBS Letters* (1976) 70:101-104.

Squires JE, Preiss T. Function and detection of 5-methylcytosine in eukaryotic RNA. *Epigenomics* (2010) 2:709-715.

Sticher, Zachary M, Gaofei Lu, Deborah G Mitchell, Joshua Marlow, Levi Moellering, Gregory R Bluemling, David B Guthrie, Michael G Natchus, George R Painter, Alexander A Kolykhalov. Analysis of the potential for N 4-hydroxycytidine to inhibit mitochondrial replication and function. *Antimicrobial Agents and Chemotherapy* (2020) 64:e01719-01719.

Suwannasom N, Kao I, Pruß A, Georgieva R, Bäumler H. Riboflavin: The health benefits of a forgotten natural vitamin. *International Journal of Molecular Sciences* (2020) 21:950.

Tanaka F, Tanaka N, Yonehara H, Umezawa H. Studies on Bacimethrin, A New Antibiotic From B. Megatherium. I Preparations and its Properties. *The Journal of Antibiotics, Series A* (1962) 15:191-196.

Tian T, Yang K-Q, Cui J-G, Zhou L-L, Zhou X-L. Folic acid supplementation for stroke prevention in patients with cardiovascular disease. *The American Journal of the Medical Sciences* (2017) 354:379-387.

Trimurtulu, Golakoti, D John Faulkner, Nigel B. Perry, Laurent Ettouati, Marc Litaudon, John W. Blunt, Murray HG Munro, Geoffrey B Jameson. Alkaloids from the antarctic sponge Kirkpatrickia varialosa. Part 2: Variolin A and N (3′)-methyl tetrahydrovariolin B. *Tetrahedron* (1994) 50:3993-4000.

Ubukata M, Morita T-I, Uramoto M, Osada H. Sparoxomycins Al and A2, New Inducers of the Flat Reversion of NRK Cells Transformed by Temperature Sensitive Rous Sarcoma Virus II. Isolation, Physico-chemical Properties and Structure Elucidation. *The Journal of Antibiotics* (1996) 49:65-70.

Wang, Haishan, Kah Leong Lim, Su Ling Yeo, Xiaoli Xu, Mui Mui Sim, Anthony E Ting, Yue Wang, Sidney Yee, YH Tan and Catherine J Pallen. Isolation of a novel protein tyrosine phosphatase inhibitor, 2-methyl-fervenulone, and its precursors from Streptomyces. *Journal of Natural Products* (2000) 63:1641-1646.

Warren R. Modified bases in bacteriophage DNAs. *Annual Reviews in Microbiology* (1980) 34:137-158.

Welch AD. Folic acid: discovery and the exciting first decade. *Perspectives in Biology and Medicine* (1983) 27:64-75.

White JD, Hansen JD. Total Synthesis of (−)-7-Epicylindrospermopsin, a Toxic Metabolite of the Freshwater Cyanobacterium Aphanizomenon o valisporum, and Assignment of Its Absolute Configuration. *The Journal of Organic Chemistry* (2005) 70:1963-1977.

Wyatt G, Cohen S. A new pyrimidine base from bacteriophage nucleic acids. *Nature* (1952) 170:1072-1073.

Xu Z, Zhang Y, Fu H, Zhong H, Hong K, Zhu W. Antifungal quinazolinones from marine-derived Bacillus cereus and their preparation. *Bioorganic & Medicinal Chemistry Letters* (2011) 21:4005-4007.

Yamaguchi H, Yamamoto C, Tanaka N. Inhibition of Protein Synthesis by Blasticidin S I. Studies with Cell-free Systems from Bacterial and Mammalian Cells. *The Journal of Biochemistry* (1965) 57:667-677.

Yoon SJ, Lyoo IK, Haws C, Kim T-S, Cohen BM, Renshaw PF. Decreased glutamate/glutamine levels may mediate cytidine's efficacy in treating bipolar depression: a longitudinal proton magnetic resonance spectroscopy study. *Neuropsychopharmacology* (2009) 34:1810-1818.

Zhang D, Miller M. Polyoxins and nikkomycins: progress in synthetic and biological studies. *Current Pharmaceutical Design* (1999) 5:73-100.

Biographical Sketch

Ravindra K. Rawal, PhD

Affiliation: Principal Scientist, CSIR-NEIST, Jorhat, Assam

Education: PhD (Chemistry), Post Doc

Business Address: CSIR- NEIST, Jorhat-785006, Assam, India.

Research and Professional Experience:
- Anti-viral Medicinal Chemistry
- Natural Product based drug discovery
- Development of Immunomodulatory formulation from natural products
- Designing of anti-HIV, anti-HBV, anti-HCV, and anti-influenza inhibitors
- Molecular modeling studies (Homology, Docking, Virtual screening and QSAR) on anti-HIV, anti-HBV, anti-HCV and anti-influenza inhibitors
- Synthesis of anti-HIV, anti-HBV, anti-HCV and anti-influenza nucleosides/nucleotide as well as non-nucleoside inhibitors

- Design and synthesis of anti-HIV prodrug
- Development of Two Clinical Candidate: FMCAP and Clevudine for HBV treatment.
- Peptide-epitope library design & synthesis

Professional Appointments:
- Principal Scientist
  CSIR-North East Institute of Science and Technology, Jorhat (August 2020-Present)
- Professor,
  Chemistry Department, Maharishi Markandeswar (Deemed to be University), Mullana (March 2018-July 2020)
- Professor, Head, Pharmaceutical Chemistry Department, Soviet College of Pharmacy, Moga (Jan 2015-March2018)
- Associate Professor, Associate Head, Pharmaceutical Chemistry and Analysis Department Indo-Soviet College of Pharmacy (Nov. 2012-Dec. 2014)
- Post Doctoral Associate, College of Pharmacy, The University of Georgia- Athens, GA, USA (Feb. 2009-Oct. 2012)
- Post Doctoral Research Fellow, City of Hope Beckmann Research Institute- Duarte, CA, USA (Jan. 2007-Feb. 2009)
- Research Fellow, Central Drug Research Institute, Lucknow, UP 226001, India (Mar. 2002-Jan. 2007)
- Chemist, Zydus-Cadila Research Centre-Ahmedabad-380015, India (Apr. 2001-Feb. 2002)

Honors:
- Selected as Bentham Ambassador 2018-2019 by Bentham Science.
- Recipient of Bharat Jyoti Award-2013 for Outstanding services, Achievement & Contribution in Education and Research at India-International Center, New Delhi
- Recipient of Certificate of Excellence-2012 for the expertise in antiviral drug discovery by University of Georgia, USA
- Recipient of Dr. M.M. Dhar Memorial Prize-2005 for Best Thesis in Chemical Sciences by Central Drug Research Institute (CDRI), Lucknow, U.P., India
- Recipient of Top Cited Author 2005-2008 Certificate from ACS National Meeting in Philadelphia, USA.

- Editorial Board Member of Archives of Organic and Inorganic Chemical Sciences (AOICS) http://lupinepublishers.us/aoics/editorial-committee.php.
- Editorial Board Member of Medicinal and Analytical Chemistry International
- Journal (MACIJ) https://medwinpublishers.com/MACIJ/editorial-board.php.
- Editorial Board Member of Open Access Journal of Pharmaceutical Research https://medwinpublishers.com/OAJPR/editorial-board.php.

Publications from the Last 3 Years:

Kiewhuo, Kikrusenuo, Dipshikha Gogoi, Hridoy Jyoti Mahanta, Ravindra K. Rawal, Debabrata Das, and G. Narahari Sastry. "North East India medicinal plants database (NEI-MPDB)." *Computational Biology and Chemistry* 100 (2022): 107728.

Jamir, Esther, Himakshi Sarma, Lipsa Priyadarsinee, Selvaraman Nagamani, Kikrusenuo Kiewhuo, Anamika Singh Gaur, Ravindra K. Rawal, Natarajan Arul Murugan, Venkatesan Subramanian, and G. Narahari Sastry. "Applying polypharmacology approach for drug repurposing for SARS-CoV2." *Journal of Chemical Sciences* 134, no. 2 (2022): 1-24.

Singh, Ranapartap, Yunes MMA Alsayadi, Vikram Jeet Singh, Pooja A. Chawla, and Ravindra K. Rawal. "Prospects of Treating Prostate Cancer through Apalutamide: A Mini-Review." *Anti-Cancer Agents in Medicinal Chemistry (Formerly Current Medicinal Chemistry-Anti-Cancer Agents)* 22, no. 6 (2022): 1056-1067.

Gupta, V., N. K. Bhardwaj, and Ravindra K. Rawal. "Removal of colour and lignin from paper mill wastewater using activated carbon from plastic mix waste." *International Journal of Environmental Science and Technology* 19.4 (2022): 2641-2658.

Pathania, Shelly, Ravindra K. Rawal, and Pankaj Kumar Singh. "RdRp (RNA-dependent RNA polymerase): A key target providing anti-virals for the management of various viral diseases." *Journal of Molecular Structure* 1250 (2022): 131756.

Kumar, Y. Bhargav, Ravindra K. Rawal, Ashutosh Thakur, and G. Narahari Sastry. "Reversible and irreversible functionalization of graphene." In *Theoretical and Computational Chemistry*, vol. 21, pp. 157-189. Elsevier, 2022.

Pathania, S., P. K. Singh, R. K. Narang, and Ravindra K. Rawal. "Structure based designing of thiazolidinone-pyrimidine derivatives as ERK2 inhibitors: Synthesis and in vitro evaluation." *SAR and QSAR in Environmental Research* 32, no. 10 (2021): 793-816.

Banerjee, Anindita, Upasana Ganguly, Sarama Saha, Suddhachitta Chakrabarti, Reena V. Saini, Ravindra K. Rawal, Luciano Saso, and Sasanka Chakrabarti. "Vitamin D and immuno-pathology of COVID-19: many interactions but uncertain therapeutic benefits." *Expert Review of Anti-infective Therapy* 19, no. 10 (2021): 1245-1258.

Bhatia, Rohit, Sankha S. Chakrabarti, Upinder Kaur, Gaurav Parashar, Anindita Banerjee, and Ravindra K. Rawal. "Multi-target directed ligands (MTDLs): promising coumarin hybrids for Alzheimer's disease." *Current Alzheimer Research* 18, no. 10 (2021): 802-830.

Vipul, Gupta, Bhardwaj Nishi Kant, and Ravindra K. Rawal. "Remediation of Chlorophenolic Compounds from Paper Mill Effluent Using High-Quality Activated Carbon from Mixed Plastic Waste." *Water, Air, & Soil Pollution* 232.8 (2021): 1-13.

Pawar, Swati, Kapil Kumar, Manish K. Gupta, and Ravindra K. Rawal. "Synthetic and medicinal perspective of fused-thiazoles as anticancer agents." *Anti-Cancer Agents in Medicinal Chemistry (Formerly Current Medicinal Chemistry-Anti-Cancer Agents)* 21, no. 11 (2021): 1379-1402.

Bhatia, Rohit, Amit Sharma, Raj K. Narang, and Ravindra K. Rawal. "Recent Nanocarrier Approaches for Targeted Drug Delivery in Cancer Therapy." *Current Molecular Pharmacology* 14, no. 3 (2021): 350-366.

Pathania, Shelly, Pankaj Kumar Singh, Raj Kumar Narang, and Ravindra K. Rawal. "Identifying novel putative ERK1/2 inhibitors via hybrid scaffold hopping–FBDD approach." *Journal of Biomolecular Structure and Dynamics* (2021): 1-16.

Singh, Ranapartap, Pooja Chawla, and Ravindra K. Rawal. "Development and Validation of Analytical Method for Simultaneous Estimation of Clidinium Bromide, Rabeprazole, Chlordiazepoxide and Dicyclomine Hydrochloride." (2021).

Karan, Ram, Rohit Bhatia, and Ravindra K. Rawal. "Applications of homogeneous catalysis in organic synthesis." In *Green Sustainable Process for Chemical and Environmental Engineering and Science*, pp. 159-188. Elsevier, 2021.

Pathania, Shelly, Parveen Bansal, Prasoon Gupta, and Ravindra K. Rawal. "Genus Calotropis: A hub of medicinally active phytoconstituents." *Current Traditional Medicine* 6, no. 4 (2020): 312-331.

Virk, Jaswinder K., Vikas Gupta, Mukesh Maithani, Ravindra K. Rawal, Sanjiv Kumar, Ranjit Singh, and Parveen Bansal. "Isolation of Sinapic Acid from Habenaria intermedia D. Don: A New Chemical Marker for the Identification of Adulteration and Substitution." *Current Traditional Medicine* 6, no. 4 (2020): 380-387.

Singh, P. K., S. Pathania, and Ravindra K. Rawal. "Exploring RdRp–remdesivir interactions to screen RdRp inhibitors for the management of novel coronavirus 2019-nCoV." *SAR and QSAR in Environmental Research* 31.11 (2020): 857-867.

Bhatia, Rohit, Subrahmanya S. Ganti, Raj Kumar Narang, and Ravindra K. Rawal. "Strategies and challenges to develop therapeutic candidates against COVID-19 pandemic." *The Open Virology Journal* 14, no. 1 (2020).

Bhatia, Rohit, Raj K. Narang, and Ravindra K. Rawal. "Coumarin-Dihydropyrimidinone Hybrids: Design, Virtual Screening, Synthesis and Cytotoxic Activity against Breast Cancer." *Journal of Advanced Scientific Research* 11 (2020).

Bhatia, Rohit, Raj K. Narang, and Ravindra K. Rawal. "In silico Investigation of Therapeutic Potentials of Coumarin-Quinoxaline Hybrids against Breast Cancer, Synthesis and In vitro Activity." *Indian Journal of Heterocyclic Chemistry* 30, no. 4 (2020): 489-502.

Chakrabarti, Sankha S., Venkatadri S. Sunder, Upinder Kaur, Sapna Bala, Priyanka Sharma, Manjari Kiran, Ravindra K. Rawal, and Sasanka Chakrabarti. "Identifying the mechanisms of α-synuclein-mediated cytotoxicity in Parkinson's disease: new insights from a bioinformatics-based approach." *Future Neurology* 15, no. 3 (2020): FNL49.

Bhatia, Rohit, Raj Kumar Narang, and Ravindra K. Rawal. "Drug repurposing-a promising tool in drug discovery against CoV-19." *Biomedical Journal of Scientific & Technical Research* 28.5 (2020): 21913-21915.

Dadwal, Ankita, Neeraj Mishra, Ravindra K. Rawal, and Raj Kumar Narang. "Development and characterisation of clobetasol propionate loaded Squarticles as a lipid nanocarrier for treatment of plaque psoriasis." *Journal of Microencapsulation* 37, no. 5 (2020): 341-354.

Kumar, Anshul, Rohit Bhatia, Pooja Chawla, Durgadas Anghore, Vipin Saini, and Ravindra K. Rawal. "Copanlisib: Novel PI3K inhibitor for the treatment of lymphoma." *Anti-Cancer Agents in Medicinal Chemistry (Formerly Current Medicinal Chemistry-Anti-Cancer Agents)* 20, no. 10 (2020): 1158-1172.

Bhatia, Rohit, Raj K. Narang, and Ravindra K. Rawal. "Repurposing of RdRp Inhibitors against SARS-CoV-2 through molecular docking tools." *Coronaviruses* 1, no. 1 (2020): 108-116.

Kaur, G., V. Gupta, P. Bansal, S. Kumar, Ravindra K. Rawal, and R. G. Singhal. "Isolation of lupenone (18-Lupen-3-one) from Roscoea purpurea root extract." *Bangladesh Journal of Medical Science* 19, no. 4 (2020): 692-696.

Pathania, Shelly, and Ravindra K. Rawal. "Green synthetic strategies toward thiazoles: a sustainable approach." *Chemistry of Heterocyclic Compounds* 56.4 (2020): 445-454.

Kumar, Kapil, and Ravindra K. Rawal. "CuI/DBU-Mediated MBH Reaction of Isatins: A Convenient Synthesis of 3-Substituted-3-hydroxy-2-oxindole." *Chemistry Select* 5.10 (2020): 3048-3051.

Manish K Gupta, Sangram K Lenka, Swati Gupta, Ravindra K Rawal, Agonist, antagonist and signaling modulators of ABA receptor for agronomic and post-harvest management. Plant Physiology and Biochemistry, 2020, 148, 10-25.

Kaur, Gunpreet, Vikas Gupta, R. G. Singhal, Ravindra K. Rawal, and Parveen Bansal. "Isolation of Catechins from Roscoea purpurea." *Journal of Young Pharmacists* 12, no. 4 (2020): 389.

Pathania Shelly, and Ravindra K. Rawal. "An update on chemical classes targeting ERK1/2 for the management of cancer." *Future Medicinal Chemistry* 12.7 (2020): 593-611.

Pathania, Shelly, Raj Kumar Narang, and Ravindra K. Rawal. "Role of sulphur-heterocycles in medicinal chemistry: An update." *European journal of medicinal chemistry* 180 (2019): 486-508.

Bhatia, Rohit, and Ravindra K. Rawal. "Coumarin hybrids: promising scaffolds in the treatment of breast cancer." *Mini Reviews in Medicinal Chemistry* 19.17 (2019): 1443-1458.

Sundeep Kaur Manjal, Shelly Pathania, Rohit Bhatia, Ramandeep Kaur, Kapil Kumar, Ravindra K. Rawal, Diversified Synthetic Strategies for Pyrroloindoles: An Overview. Journal of Heterocyclic Chemistry, 2019, 56(9), 2318-2332.

R Raturi, M Maithani, V Gupta, RG Singhal, Ravindra K Rawal, S Kumar, P Singh, R and Bansal, Isolation of 5-o-caffeoylquinic acid from natural source microstylis muscifera (lindl.) Kuntze: a potential chemical marker for identification. *International Journal of Recent Scientific Research*, 2019, 10(8), 34456-34459.

R Raturi, M Maithani, V Gupta, RG Singhal, Ravindra K. Rawal, S Kumar, R Singh, P Bansal, First Report on Isolation of N-Isopropyl Palmitamide from the Natural Source Microstylis muscifera: a Potential Chemical Marker for Identification. *Chemistry of Natural Compounds,* 2019, 55(4), 773-774.

Ramandeep Kaur, Yagyesh Kapoor, Sundeep Kaur Manjal, Ravindra K. Rawal And Kapil Kumar, Diversity-Oriented Synthetic Approaches For Furoindoline: A Review. *Current Organic Synthesis* 2019, 16, 342-368.

Chapter 3

Regulation of the Pyrimidine Biosynthetic Pathway in the Bacterium *Pseudomonas chlororaphis*

**Akram Bani Ahmad
and Thomas P. West**[*]

Department of Chemistry, Texas A&M University-Commerce, Commerce, Texas, USA

Abstract

The pyrimidine biosynthetic pathway consists of five enzymes that are unique to the formation of pyrimidine nucleotides. The pyrimidine biosynthetic pathway enzymes are aspartate transcarbamoylase, dihydroorotase, dihydroorotate dehydrogenase, orotate phosphoribosyltransferase and orotidine 5'-monophosphate decarboxylase. The focus of this study was to explore the regulation of pyrimidine biosynthesis in *Pseudomonas chlororaphis* ATCC 17414. To do so, the effect of supplementing the pyrimidine bases orotic acid or uracil into a glucose or succinate-containing culture medium of *P. chlororaphis* ATCC 17414 on the pyrimidine biosynthetic enzyme activities was first explored. Transcarbamoylase and decarboxylase activities were repressed by orotic acid or uracil addition independent of carbon source in the ATCC 17414 cells. Next, a pyrimidine auxotrophic mutant of *P. chlororaphis* was isolated using chemical mutagenesis and 5-fluororotic acid resistance. The isolated mutant utilized either uracil, uridine or cytosine as a pyrimidine source to support their growth. The mutant strain was deficient for orotidine 5'-monophosphate decarboxylase activity. The mutant cells were subjected to pyrimidine

---

[*] Corresponding Author's Email: Thomas.West@tamuc.edu.

In: Pyrimidines and Their Importance
Editor: Roger G. Ward
ISBN: 979-8-88697-656-4
© 2023 Nova Science Publishers, Inc.

limitation to determine if nucleotide depletion influenced the synthesis of the pyrimidine biosynthetic enzyme activities. The pyrimidine limitation of a pyrimidine auxotrophic strain often causes derepression of the synthesis of the pyrimidine pathway enzymes. When the orotidine 5'-monophosphate decarboxylase mutant strain cells were limited for pyrimidines for one hour, transcarbamoylase, dihydroorotase, dehydrogenase and phosphoribosyltransferase activities were all derepressed compared to their activities in the mutant strain grown in excess uracil. To learn if the initial pathway enzyme aspartate transcarbamoylase was subject to regulation of its activity, the effect of pyrophosphate and ribonucleotides in *P. chlororaphis* ATCC 17414 cells was studied. The transcarbamoylase was highly inhibited by uridine 5'-monophosphate, uridine 5'-diphosphate, cytidine 5'-monophosphate and guanosine 5'-triphosphate in the glucose-grown ATCC 17414 cells. Overall, it was concluded that pyrimidine biosynthetic pathway enzymes in *P. chlororaphis* ATCC 17414 was regulated at the level of enzyme synthesis and that the initial pathway enzyme aspartate transcarbamoylase activity was controlled at the level of enzyme activity.

Keywords: pyrimidine, biosynthesis, bacterium, pseudomonad, orotic acid, uracil

Introduction

There are five enzymes unique to the *de novo* pyrimidine biosynthetic pathway in pseudomonads (Chu and West, 1990a; West, 1994a; West, 1997a; Haugaard and West, 2002; Santiago and West, 2002a; Santiago and West, 2002b; West, 2002; West, 2004a; West 2004b; West, 2004c; West, 2005a; West, 2005b; West, 2007a; West; 2007b; West, 2009; West, 2010a; West, 2012; West, 2014; Chunduru and West, 2018; Murahari and West, 2019; Domakonda and West, 2020). The first enzyme is aspartate transcarbamoylase which catalyzes the reaction of carbamoyl phosphate and aspartate to form *N*-carbamoyl L-aspartate along with inorganic phosphate (Pardee and Yates, 1956a; Pardee and Yates, 1956b). Transcarbamoylase is known to be highly regulated at the level of enzyme activity in a number of pseudomonads (Condon et al., 1976; West, 1997b; Santiago and West, 2003a). Cyclization of *N*-carbamoyl L-aspartate cyclization is catalyzed by dihydroorotase where elimination of a water molecule allows the cyclization of carbamoyl aspartate to form dihydroorotate. Dihydroorotate is then oxidized by a derivative of coenzyme Q to form orotate in a reaction catalyzed by the third enzyme involved in the pyrimidine biosynthetic pathway, namely dihydroorotate dehydrogenase. The addition of ribose 5-phosphate occurs using 5'-phosphoribosyl 1-pyrophosphate where orotate is bonded to ribose 5-phosphate to form orotidine 5'-monophosphate in an irreversible step catalyzed by orotate phosphoribosyltransferase. The last step in the biosynthetic pathway is catalyzed by the fifth pathway enzyme, orotidine 5'-monophosphate decarboxylase which involves the irreversible decarboxylation of orotidine 5'-monophosphate to uridine 5'- monophosphate. Uridine 5'- monophosphate is considered the precursor of all pyrimidine-based nucleotides like uridine 5'-triphosphate and cytidine 5'-triphosphate. For instance, uridine 5'-monophosphate is phosphorylated to form uridine 5'-diphosphate and uridine 5'-triphosphate in reactions catalyzed by kinases. The regulatory enzyme cytidine 5'-triphosphate synthetase is responsible for producing cytidine 5'-triphosphate from its substrates uridine 5'-triphosphate and glutamine (West and O'Donovan, 1982; West et al., 1983). This enzyme is necessary for the synthesis of ribonucleic acid. Subsequently, the ribonucleotides can be used to form deoxyribonucleotides vital for deoxyribonucleic acid formation (O'Donovan and Neuhard, 1970). The degradation products of ribonucleic acid, which include the pyrimidine bases cytosine and uracil, can be further degraded in pseudomonads to provide a source of nitrogen (Kim and West, 1991; West, 1992; Xu and West, 1992; West, 1994b; West, 1996; Santiago and West, 1999; West, 2000; West, 2001; Gill and West, 2022).

The gram-negative bacterium *Pseudomonas chlororaphis* is present in a variety of habitats such as in soil or freshwater ecosystems. This pseudomonad is also known to be associated with plants. It has been shown that *P. chlororaphis* can serve as a biofertilizer and has antifungal activity. Considering its biological properties, *P. chlororaphis* has numerous applications in agriculture (Patten and Glick, 2002; Pohanka et al., 2005; Mehnaz et al., 2010). This microbe stimulates plant growth by promoting rhizobacterial growth that limits the chance of disease occurrence. It has also been demonstrated that *P. chlororaphis* has the ability to produce phenazine-type antibiotics (Price-Whelan et al., 2006; Dietrich et al., 2006; Morohoshi et al., 2013). In particular, this bacterium has many applications in agriculture as a biocontrol agent against particular fungal plant pathogens (Chin et al., 2005; Mehnaz et al., 2013; Mehnaz et al., 2014). The antibiotics synthesized by *P. chlororaphis* have been noted to promote antifungal efficacy against such fungal species as *Rhizoctonia solani, Botrytis cinerea, Verticillium dahlia,* and *Sclerotinia sclerotiorum.*

The taxonomy of *P. chlororaphis* has been investigated and it has been classified within the *Pseudomonas chlororaphis* homology group based on 16S ribosomal ribonucleic acid analysis (Anzai et al., 2000). The pseudomonad species classified in this homology group include *P. chlororaphis, Pseudomonas aurantiaca, Pseudomonas fragi, Pseudomonas lundensis* and *Pseudomonas taetrolens* (Anzai et al., 2000; Mulet et al., 2010). The *de novo* pyrimidine biosynthetic pathways in *P. aurantiaca, P. fragi, P. lundensis* and *P. taetrolens* have been shown to be regulated by pyrimidines at the level of enzyme synthesis (West, 2002; West, 2004a; West 2009; Domakonda and West, 2020). In addition, *in vitro* regulation of transcarbamoylase activity in these species by pyrophosphate and ribonucleotides has been shown to exist (West, 2002; West, 2004a; West 2009; Domakonda and West, 2020). Therefore, the focus of this investigation was to determine whether the pyrimidine biosynthetic pathway enzymes in *P. chlororaphis* were also subject to transcriptional regulation and whether its aspartate transcarbmoylase activity was controlled by pyrophosphate and pyrimdine ribonucleotides.

Methods

Bacterial Strain and Growth Conditions

The bacterial strain in this study was *Pseudomonas chlororaphis* ATCC 17414 which was obtained from the American Type Culture Collection (Manassas, Virginia, USA). The strain was preserved on a solid nutrient agar medium and solid minimal media. The minimal medium was comprised of 7.4 mM potassium phosphate dibasic, 5.7 mM potassium phosphate monobasic medium, 17.1 mM sodium chloride, 1.7 mM sodium citrate dihydrate, 28.4 mM magnesium sulfate heptahydrate and 30.3 mM ammonium sulfate (Stanier, 1947; West, 1989). After preparation, the medium pH was adjusted to 7.0-7.2 prior to sterilization. Glucose or sodium succinate (0.4%, w/v) was added to the medium as a carbon source. If a pyrimidine base was supplemented into the medium, it was added at a concentration of 50 mg/L. To make solid medium, agar (2%, w/v) was added. To sterilize the media, it was sterilized for 20 min at 121°C and 20 pounds/square inch using an autoclave. Liquid cultures were shaken on a rotary shaker (200 revolutions/minute) at 30°C to provide adequate aeration. In the pyrimidine limitation experiments using the auxotrophic strain, cultures of liquid medium containing 50 mg/L uracil and either 0.4% glucose or succinate as the carbon source were grown until the late exponential phase of growth was reached. The cells were collected using centrifugation at 7,719 x g for 20 minutes at 4°C under sterile conditions and washed with sterile minimal medium lacking uracil and then resuspended in the respective minimal medium minus uracil. The cultures were aerated on a rotary shaker (200 evolutions/minute) at 30°C for a period of 1 or 2 hours of pyrimidine limitation and the cells were collected and the cell extracts prepared.

Isolation and Characterization of a Pyrimidine Auxotroph

The procedure to isolate a pyrimidine auxotroph involved ethyl methanesulfonate mutagenesis of the ATCC 17414 cells and testing the mutagenized cells for resistance to the pyrimidine analogue 5-fluoroorotic acid (Watson and Holloway, 1976; Watrin et al., 1999; Santiago and West, 2002a). An exponentially-growing *P. chlororaphis* nutrient broth (5 mL) culture was treated with 1% ethyl methanesulfonate for 90 minutes at 30°C without shaking (Santiago and West, 2002b). Subsequently, the treated culture was diluted (1:30) in glucose minimal medium comprising (50 mg/L) uracil. The culture was shaken overnight at 30°C to permit outgrowth of any potential mutants. Mutagenized cells were spread on glucose minimal medium agar plates having 50 mg/L 5-fluoroorotic acid and 1 mg/L uracil. The tiniest colonies growing on the solid medium agar plates containing the pyrimidine analogue were picked and screened for their ability to grow on glucose minimal medium agar plates and glucose minimal medium agar plates containing 50 mg/L uracil. *P. chlororaphis* strain BW-1 was identified as uracil-requiring strain. To determine the auxotrophic requirements of mutant strain BW-1, about 10^8 cells of the mutant strain were spread onto glucose or succinate minimal medium agar plates. A glass micro-fiber filter disk (2 cm diameter) saturated with a sterile solution of *N*-carbamoyl L-aspartate, dihydroorotic acid, orotic acid, uracil, cytosine, uridine, cytidine, dihydrouracil, cytidine 5'-monophosphate or uridine 5'-monophosphate was placed at the center of each inoculated plate. The plates were then incubated at 30°C for an eight-day period and examined daily. Any confluent growth surrounding the disk on each plate was recorded as positive for pyrimidine utilization.

Cell Extraction Preparation

For cell extract preparation, inoculated liquid medium cultures (40 mL) were incubated at 30°C and aerated by shaking at 200 revolutions/minute (West, 2005a). Growth of the cultures was monitored spectrophotometrically at 600 nanometers. When growth reached the late exponential phase of growth or after the mutant strain was limited for pyrimidines for 1 or 2 hours, the cells were collected by being centrifuged at 7,719 x g for 20 minutes at 4°C. After collection, cells were washed with 0.85% NaCl (25 mL) and centrifuged at 7,719 x g for 20 minutes at 4°C. The cell pellet was then suspended in 20 mM Tris-HCl buffer (pH 8.5), 1 mM ethylene diaminetetracetic acid and 1 mM 2-mercaptoethanol (2.5 mL). To disrupt the bacterial cell membranes, the cell suspension was sonicated at the highest power for the ultrasonic microtip for eight times using burst of 30 seconds while cooling the sample in an ice bath. Following the sonication procedure, the cell extracts were centrifuged at 1,930 x g for 15 minutes at 4°C and then dialyzed for 18 hours against 300 mL of 20 mM Tris-hydrochloride buffer (pH 8.0) and 1 mM 2-mercaptoethanol at 4°C. Following dialysis, cell extracts were assayed for the five *de novo* pyrimidine biosynthetic enzyme activities.

Enzymes Assays

The *de novo* pyrimidine biosynthetic pathway were assayed spectrophotometrically at 30°C (Beckwith et al., 1962; Prescott and Jones, 1969; Adair and Jones, 1972; Schwartz and Neuhard, 1975; Kelln et al., 1975; Haugaard and West, 2002; West, 2014). Aspartate transcarbamoylase activity was measured over a 30 minute period using an assay mix (1 mL) consisting of 0.1 M Tris-HCl buffer (pH 8.5), 10 mM L-aspartic acid (pH 8.5), 1 mM carbamoyl phosphate and cell extract (Adair and Jones, 1972). The reaction was initiated by carbamoyl phosphate (1 mM) addition and the reaction was halted at selected time intervals by adding color mix (1 mL) to the assay tubes. The color mix reagent was freshly prepared before use and consisted of two parts to one part of antipyrine reagent and oxime reagent (Prescott and Jones, 1969). Antipyrine reagent consisted of 5 g of antipyrine (2,3-dimethyl-1-phenyl-3-pyrazine) dissolved in a liter of 50% (v/v) sulfuric acid (Prescott and Jones, 1969). The oxime reagent was composed of 0.8 g 2,3-butanedione monoxime dissolved in 5% (v/v) acetic acid (Prescott and Jones, 1969). Similarly, dihydroorotase was also assayed using a colorimetric method. The modified assay mixture (1 mL) contained 0.1 M Tris-HCl buffer (pH 8.5), 1 mM EDTA, 1 mM dihydroorotate in 100 mM potassium phosphate buffer (pH 7.0) and cell extract (Beckwith et al., 1962). The reaction was started using L-dihydroorotate and its enzyme activity was assayed over 30 minutes at selected intervals with the reaction being stopped using 1 mL of color mix. After color mix addition, the transcarbamoylase and dihydroorotase assay tubes were incubated at 60°C under fluorescent lighting for 2 hours (Prescott and Jones, 1969). The assay tubes were cooled to room temperature and their absorbance were measured at 466 nanometers. Transcarbamoylase and dihydroorotase activities were calculated using an experimentally determined absorption coefficient derived from a standard curve at 466 nanometers plotted against selected concentrations of N-carbamoyl L-aspartic acid (Prescott and Jones, 1969). A unit of aspartate transcarbamoylase activity was given as nanomoles of N-carbamoyl L-aspartic acid produced per minute at 30°C while a unit of dihydroorotase activity was expressed as nanomoles of N-carbamoyl L-aspartate utilized per minute at 30°C. When determining the Michaelis-Menten constant (K_m) of aspartate transcarbamoylase for its substrates carbamoyl phosphate and L-aspartic acid, the L-aspartic acid or carbamoyl phosphate concentration was varied in the assay mixture depending upon which substrate was being studied (West, 1997b). The next three biosynthetic enzymes were assayed by continuous spectrophotometric procedures (Chunduru and West, 2018). Dihydroorotate dehydrogenase was assayed spectrophotometrically using a modified assay mixture (1 mL). The modified assay mixture consisted of 0.1 M Tris-HCl buffer (pH 8.5), 2 mM dihydroorotic acid and cell extract. The basis for this assay is the difference in the extinction coefficients between dihydroorotate and orotate at 290 nanometers and pH 8.5 which equals 6.5×10^3 M^{-1} cm^{-1}. A unit of dihydroorotate dehydrogenase activity was given as nanomoles of orotate formed/minute at 30°C. Orotate phosphoribosyltransferase enzyme activity was measured in a reaction mixture (1 mL) containing 0.1 M Tris-HCl buffer (pH 8.8), 6 mM $MgCl_2$ and cell extract. The rection was initiated by 5 mM orotic acid and 6 mM 5-phosphoribosyl 1-pyrophosphate addition and the absorbance at 295 nanometers of the reaction was followed spectrophotometrically. The difference in the extinction coefficients at 295 nm between orotate and OMP at pH 8.8 (3.67×10^3 M^{-1} cm^{-1}) allowed the reaction to be studied spectrophotometrically. A unit of orotate phosphoribosyltransferase activity was stated as nanomoles of orotidine 5'-monophosphate formed/minute at 30°C. Orotidine 5'-monophosphate decarboxylase activity was assayed spectrophotometrically at 290 nanometers where the reaction mixture (1 mL) contained 0.1 M Tris-HCl buffer (pH 8.8), 6 mM $MgCl_2$ and cell extract. The reaction was initiated by orotidine 5'-monophosphate addition to the mixture. The difference in extinction coefficients at 290 nanometers (1.38 x 10^3 M^{-1} cm^{-1}) between orotidine 5'-monophosphate and uridine 5'-monophosphate at pH 8.5 was used to measure its enzyme activity. A unit of decarboxylase activity was expressed as nanomoles of uridine 5'-monophosphate formed/minute at 30°C.

Protein Assay, Specific Activity and Statistical Analysis

Protein concentration was determined by the method of Bradford (1976). The concentration of protein in each sample was calculated using a standard curve in which bovine serum albumin served as the standard. The specific activity of each pyrimidine biosynthetic enzyme was expressed as nmoles product/minute/milligram protein. The data gathered during the effect of pyrimidine supplementation studies and the pyrimidine limitation experiments represents the mean of three separate determinations using three independent cultures. The values used during the aspartate transcarbamoylase regulation and kinetics experiments were the average of two separate trials.

Results

Initially, the presence of the pyrimidine biosynthetic pathway enzyme activities in *P. chlororaphis* ATCC 17414 cells was investigated. When *P. chlororaphis* ATCC 17414 cells were grown in media containing glucose as the carbon source, the five *de novo* pyrimidine biosynthetic enzymes were found to be active (Table 1). Of the five *de novo* pathway enzymes assayed, aspartate transcarbamoylase had the highest activity while dihydroorotase dehydrogenase exhibited the lowest activity (Table 1). Similarly, it was observed that the *de novo* pyrimidine biosynthetic pathway enzymes were active in the *P. chlororaphis* ATCC 17414 cells grown on succinate as a carbon source (Table 2). Dihydroorotase had the highest activity in the succinate-grown ATCC 17414 cells while dihydroorotate dehydrogenase exhibited the lowest activity (Table 2). The *de novo* pyrimidine biosynthetic

Table 1. The effect of pyrimidine base supplementation on the *de novo* pyrimidine biosynthetic enzyme activities in glucose-grown cells of *Pseudomonas chlororaphis* ATCC 17414

Enzyme	Pyrimidine supplement	Specific activity (standard deviation)
Aspartate transcarbamoylase	None	45.6 (1.4)
Aspartate transcarbamoylase	Orotic acid	22.3 (0.8)
Aspartate transcarbamoylase	Uracil	11.6 (1.2)
Dihydroorotase	None	17.3 (0.5)
Dihydroorotase	Orotic acid	67.8 (1.2)
Dihydroorotase	Uracil	12.9 (1.9)
Dihydroorotate dehydrogenase	None	16.9 (1.1)
Dihydroorotate dehydrogenase	Orotic acid	7.5 (0.9)
Dihydroorotate dehydrogenase	Uracil	19.0 (0.7)
Orotate phosphoribosyltransferase	None	29.1 (1.6)
Orotate phosphoribosyltransferase	Orotic acid	38.0 (1.5)
Orotate phosphoribosyltransferase	Uracil	55.2 (1.1)
Orotidine 5'-monophosphate decarboxylase	None	12.5 (1.3)
Orotidine 5'-monophosphate decarboxylase	Orotic acid	6.1 (1.3)
Orotidine 5'-monophosphate decarboxylase	Uracil	7.4 (0.1)

Table 2. The effect of pyrimidine base supplementation on the *de novo* pyrimidine biosynthetic enzyme activities in succinate-grown cells of *Pseudomonas chlororaphis* ATCC 17414

Enzyme	Pyrimidine supplement	Specific activity (standard deviation)
Aspartate transcarbamoylase	None	12.5 (1.3)
Aspartate transcarbamoylase	Orotic acid	9.2 (0.9)
Aspartate transcarbamoylase	Uracil	14.0 (0.5)
Dihydroorotase	None	75.0 (4.7)
Dihydroorotase	Orotic acid	37.5 (7.0)
Dihydroorotase	Uracil	42.9 (0.3)
Dihydroorotate dehydrogenase	None	7.4 (0.1)
Dihydroorotate dehydrogenase	Orotic acid	6.7 (0.6)
Dihydroorotate dehydrogenase	Uracil	6.6 (0.4)
Orotate phosphoribosyltransferase	None	22.5 (0.8)
Orotate phosphoribosyltransferase	Orotic acid	24.5 (1.2)
Orotate phosphoribosyltransferase	Uracil	18.0 (0.7)
Orotidine 5'-monophosphate decarboxylase	None	35.0 (0.5)
Orotidine 5'-monophosphate decarboxylase	Orotic acid	22.5 (2.2)
Orotidine 5'-monophosphate decarboxylase	Uracil	22.8 (0.6)

enzyme activities were generally lower in the succinate-grown *P. chlororaphis* cells (Table 2) than the activities determined in the glucose-grown cells (Table 1). It should be mentioned that carbon source has been shown to influence the pyrimidine pyrimidine biosynthetic enzyme activities in other pseudomonads (Santiago and West, 2003b; Santiago and West, 2003c; West, 2005c; West, 2010b).

The influence of pyrimidine base supplementation in the minimal medium cultures of ATCC 17414 on the pyrimidine biosynthetic pathway enzyme activities was investigated. More specifically, the supplementation of orotic acid or uracil (50 mg/L) to minimal medium cultures of ATCC 17414 was performed to determine if the nucleotide metabolites of orotic acid or uracil could repress the levels of the pyrimidine biosynthetic pathway enzymes. When glucose served as the carbon source, uracil addition decreased the activity of aspartate transcarbamoylase by 50% in *P. chlororaphis* cells grown on glucose as a carbon source (Table 1). Orotic acid supplementation decreased the activity in the glucose-grown ATCC 17414 cells by 70% relative to the unsupplemented cells (Table 1). Uracil supplementation to the succinate minimal medium of the ATCC 17414 cells resulted in a 30% decrease in aspartate transcarbamoylase activity compared to the unsupplemented cells (Table 1). A slight increase in the transcarbamoylase activity occurred when the succinate-grown cells were grown with orotic acid (Table 2). For the glucose-grown ATCC 17414 cells, dihydroorotase activity decreased slightly following uracil addition to the minimal medium relative to the unsupplemented glucose-grown cells (Table 1). There was a highly significant elevation (3.9-fold) in dihydroorotase activity in the ATCC 17414 cells following orotic acid supplmentation (Table 1). A 50% or 40% decrease in dihydroorotase activity was observed following orotic acid or uracil inclusion in the medium, respectively, in the succinate-grown ATCC 17414 cells (Table 2). Dihydroorotate dehydrogenase activity decreased by 60% in the glucose-grown ATCC 17414 cells when uracil was present in the medium while dehydrogenase activity was affected slightly after orotic acid supplementation compared to unsupplemented cells (Table 1). In succinate-grown cells of *P. chlororaphis* supplemented with uracil or orotic acid, dehydrogenase activity decreased slightly compared to the unsupplemented ATCC 17414 cells (Table 2). In contrast, a slight rise in orotate phosphoribosyltransferase activity was observed in the glucose-grown ATCC 17414 cells following uracil or orotic acid addition to the medium (Table 1). In the succinate-grown ATCC 17414 cells, uracil addition caused a 20% decrease in orotate phosphoribosyltransferase relative to the unsupplemented succinate-grown cells (Table 2). On the other hand, phosphoribosyltransferase activity increased slightly in the orotic acid-containing medium compared to the unsupplemented cells (Table 2). When glucose served as the carbon source, orotidine 5'-monophosphate decarboxylase activity diminished by 40% or 50% in the glucose-grown cells when the medium contained uracil or orotic acid, respectively (Table 1). Decarboxylase activity also dropped when *P. chlororaphis* cells were grown in succinate minimal medium containing uracil or orotic acid in comparison with unsupplemented cells (Table 2).

With the influence of pyrimidine base supplementation on the pyrimidine biosynthetic pathway activities in *P. chlororaphis* having been determined, the effect of pyrimidine limitation on a pyrimidine biosynthetic pathway mutant strain was the next objective of this study in order to see if derepression of pathway enzyme synthesis was possible to observe (Lazzarini et al., 1969). The isolation of a pyrimidine auxotrophic strain is necessary to be able to perform pyrimidine limitation experiments. Chemical mutagenesis and resistance to the the pyrimidine analogue 5-fluoroorotic acid (Watrin et al., 1999; Santiago and West, 2002a) was used to identify an orotidine 5'-monophosphate decarboxylase mutant strain of *P. chlororaphis* ATCC 17414 that was designated as strain BW-1 (Tables 3 and 4). The decarboxylase activity in glucose-grown or succinate-grown cells of strain BW-1 was not detectable (Table 3 and 4). The growth requirements for pyrimidines and related compounds was analyzed using glucose or succinate as a carbon source. Mutant strain BW-1 had the ability to utilize uracil, cytosine or uridine to meet its pyrimidine requirement regardless of whether the carbon source was glucose or succinate. The ability of mutant strain BW-1 to utilize cytosine or uridine as a pyrimidine source is likely related to the presence of cytosine deaminase or nucleoside hydrolase in *P. chlororaphis* ATCC 17414 cells similar to related *Pseudomonas* species (West, 1988; Chu and West, 1990b; West, 1991; Beck and O'Donovan, 2008).

The pyrimidine nucleotide limitation studies were initiated using strain BW-1 where the mutant strain was starved for pyrimidines for 1 or 2 hours (Condon et al. 1976; West, 2004c; West, 2007b; Schultheisz et al. 2011). It should be noted that pyrimidine limitation of strain BW-1 beyond 2 hours resulted in significant cell lysis preventing assaying of the pathway activities. In the glucose-grown mutant strain BW-1 cells, aspartate transcarbamoylase activity increased by 4.6-fold after 1 hour of pyrimidine limitation but a significant depression in its activity was observed after 2 hours of pyrimidine limitation compared to the uracil-grown mutant strain cells (Table 3). In the succinate-grown mutant cells, transcarbamoylase activity was elevated by

8.3-fold or 5.9-fold following pyrimidine limitation for 1 or 2 hours, respectively, compared to the cells grown in excess uracil (Table 4). In the glucose-grown strain BW-1 cells, pyrimidine nucleotide limitation for 1 hour increased dihydroorotase activity by 4.5-fold but 2 hours of pyrimidine limitation of the mutant cells resulted in a significant depression in dihydroorotase activity (Table 3). The succinate-grown strain BW-1 cells were shown to contain a 4.4-fold higher dihydroorotase activity after 1 hour of pyrimidine starvation and its activity further increased to 4.5-fold higher following 2 hours of pyrimidine-limiting condition compared to cells grown under saturating uracil conditions (Table 3). Dihydroorotate dehydrogenase activity in the glucose-grown following 2 hours of pyrimidine limiting condition compared to cells grown under excess uracil conditions (Table 3). Dihydroorotate dehydrogenase activity in glucose-grown strain BW-1 cells subjected to pyrimidine-limiting conditions for 1 hour increased by 8.2-fold while pyrimidine limitation for 2 hours significantly decreased dehydrogenase activity relative to mutant cells grown under saturating uracil conditions (Table 3). In the succinate-grown strain BW-1 cells, dehydrogenase activity increased after pyrimidine limitation for 1 or 2 hours by 17.8-fold or 12.6-fold, respectively, compared to strain BW-1 cells grown medium containing excess uracil (Table 4). Orotate phosphoribosyltransferase activity increased by

Table 3. The effect of pyrimidine limitation on the pyrimidine biosynthetic enzyme activities in glucose-grown cells of *Pseudomonas chlororaphis* strain BW-1

Enzyme	Pyrimidine limitation	Specific activity (standard deviation)
Aspartate transcarbamoylase	None	17.5 (2.9)
Aspartate transcarbamoylase	1 h	79.9 (4.8)
Aspartate transcarbamoylase	2 h	3.0 (0.8)
Dihydroorotase	None	21.1 (5.4)
Dihydroorotase	1 h	95.6 (1.9)
Dihydroorotase	2 h	4.1 (1.7)
Dihydroorotate dehydrogenase	None	16.4 (1.0)
Dihydroorotate dehydrogenase	1 h	134.4 (1.3)
Dihydroorotate dehydrogenase	2 h	4.2 (0.4)
Orotate phosphoribosyltransferase	None	41.3 (0.7)
Orotate phosphoribosyltransferase	1 h	75.1 (2.7)
Orotate phosphoribosyltransferase	2 h	25.6 (0.5)
Orotidine 5'-monophosphate decarboxylase	None	<1.4 (0.8)

Table 4. Influence of pyrimidine limitation on the pyrimidine biosynthetic enzyme activities in succinate-grown cells of *Pseudomonas chlororaphis* strain BW-1

Enzyme	Pyrimidine limitation	Specific activity (standard deviation)
Aspartate transcarbamoylase	None	19.4 (0.6)
Aspartate transcarbamoylase	1 h	160.0 (7.0)
Aspartate transcarbamoylase	2 h	115.1 (1.9)
Dihydroorotase	None	26.1 (0.9)
Dihydroorotase	1 h	114.1 (3.8)
Dihydroorotase	2 h	116.7 (2.9)
Dihydroorotate dehydrogenase	None	13.8 (0.3)
Dihydroorotate dehydrogenase	1 h	244.9 (0.7)
Dihydroorotate dehydrogenase	2 h	173.5 (1.7)
Orotate phosphoribosyltransferase	None	33.9 (1.0)
Orotate phosphoribosyltransferase	1 h	65.1 (2.3)
Orotate phosphoribosyltransferase	2 h	65.4 (1.0)
Orotidine 5'-monophosphate decarboxylase	None	<0.3 (0.0)

1.8-fold after pyrimidine limitation for 1 hour in glucose-grown strain BW-1 cells but its activity decreased following 2 hours of limitation in relation to its activity in glucose-grown strain BW-1 cells grown in medium that contained a saturating concentration of uracil (Table 4). In contrast, phosphoribosyltransferase activity increased by 1.9-fold after 1 or 2 hours of pyrimidine limitation in the succinate-grown strain BW-1 cells (Table 4). The pyrimidine limitation experiments utilizing strain BW-1 does provide evidence that pyrimidine biosynthetic pathway enzyme synthesis is subject to control by the concentration of pyrimidine nucleotides within the *P. chlororaphis* cell independent of whether glucose or succinate served as the carbon source.

It was also of interest to learn whether the initial pyimidine biosynthetic pathway enzyme aspartate transcarbamoylase was regulated in *P. chlororaphis* since it is known to be subject to feedback inhibition in a number of species including species of *Pseudomonas* (O'Donovan and Neuhard, 1970). It has been shown that aspartate transcarbamoylase regulates the rate of bacterial pyrimidine nucleotide formation in bacteria considering that the enzyme catalyzes the first committed step in pyrimidine biosynthesis (O'Donovan and Neuhard, 1970). Prior to examining the regulation of the *P. chlororaphis* transcarbamoylase, it was necessary to derive the Michaelis-Menten (K_m) constants for its substrates carbamoyl phosphate and L-aspartic acid. The K_m of the *P. chlororaphis* transcarbamoylase for carbamoyl phosphate or L-aspartic acid was 0.08 mM or 3.67 mM, respectively. The effect of 5 mM pyrophosphate and pyrimidine ribonucleotides on the *P. chlororaphis* transcarbamoylase activity in cell extracts was examined under saturating substrate concentrations (1 mM carbamoyl phosphate and 10 mM L-aspartic acid). Pyrophosphate highly inhibited the *P. chlororaphis* transcarbamoylase activity in ATCC 17414 glucose-grown cell extracts (Table 5). Next, pyrimidine and purine 5'-monophosphate nucleotides were examined as possible effectors of the *P. chlororaphis* transcarbamoylase activity in glucose grown cell extracts (Table 5). With respect to the ribonucleotide 5'-monophosphates tested, uridine 5'-monophosphate was the most inhibitory of transcarbamoylase activity in ATCC 17414 while the least inhibitory ribonucleotide was guanosine 5'-monophosphate (Table 5). Relative to the pyrimidine and purine ribonucleotide 5'-diphosphates screened, the most effective inhibitors of ATCC 17414 transcarbamoylase activity were uridine 5'-diphosphate and cytidine 5'-diphosphate while the least effective inhibitors of its activity were adenosine 5'-diphosphate and guanosine 5'-diphosphate (Table 5). The pyrimidine and purine ribonucleotide 5'-triphosphates were also highly inhibitory of the transcarbamoylase activity in the *P. chlororaphis* ATCC 17414 cell extracts (Table 5). Of the ribonucleotide 5'-triphosphates, the most effective inhibitor of aspartate transcarbamoylase activity was guanosine 5'-triphosphate since it reduced transcarbamoylase activity by 98% compared to the activity in the control cell extract (Table 5). In addition, transcarbamoylase activity was reduced by 94%, 89% or 60% in the presence of cytidine 5'-triphosphate, adenosine 5'-triphosphate or uridine 5'-triphosphate, respectively, when included in the reaction mixture (Table 5). Considering all the effectors tested, it was determined that uridine 5'-monophosphate or cytidine 5'-monophosphate was the most inhibitory effectors of the *P. chlororaphis* aspartate transcarbamoylase activity although uridine 5'-diphosphate or guanosine 5'-triphosphate were also highly inhibitory of the transcarbamoylase activity (Table 5). Evaluation of the findings clearly showed that aspartate transcarbamoylase activity in the *P. chlororaphis* ATCC 17414 cell extracts was strongly inhibited by pyrophosphate or ribonucleotides at the monophosphate, diphosphate or triphosphate level. It appeared that the uridine riboncleotides strongly influenced that transcarbamoylase activity in the glucose-grown cell extracts (Table 5).

Table 5. The effect of pyrophosphate and pyrimidine ribonucleotides on aspartate transcarbamoylase activity in glucose-grown cells of *Pseudomonas chlororaphis* ATCC 17414

Enzyme Effector	Specific activity (average of 2 trials)	Relative activity
None	45.6	100
Pyrophosphate	6.8	15
Guanosine 5'-monophosphate	19.0	42
Adenosine 5'-monophosphate	18.0	40
Uridine 5'-monophosphate	0.1	0
Cytidine 5'-monophosphate	<0.2	0
Guanosine 5'-diphosphate	2.5	6
Adenosine 5'-diphosphate	5.4	12
Uridine 5'-diphosphate	0.7	1
Cytidine 5'-diphosphate	1.6	4
Guanosine 5'-triphosphate	0.7	2
Adenosine 5'-triphosphate	5.1	11
Uridine 5'-triphosphate	18.2	40
Cytidine 5'-triphosphate	2.7	6

Discussion

It is important to study the regulation of pyrimidine biosynthesis in *P. chlororaphis* ATCC 17414 by pyrimidine-related compounds for several reasons. One reason is that the findings of this investiagtion could provide more credence to the current taxonomic assignment of *P. chlororaphis* to the *P. chlororaphis* homology group along with the closely-related species *P. aurantiaca, P. fragi, P. lundensis* and *P. taetrolens* (Anzai et al., 2000). It is helpful that regulation of pyrimidine biosynthesis by pyrimidine-related compounds in pseudomonads has been examined in several prior studies (Chu and West, 1990a; West, 1994a; West, 1997a; Haugaard and West, 2002; Santiago and West, 2002a; Santiago and West, 2002b; West, 2002; West, 2004a; West 2004b; West, 2004c; West, 2005a; West, 2005b; West, 2007a; West; 2007b; West, 2009; West, 2010a; West, 2012; West, 2014; Chunduru and West, 2018; Murahari and West, 2019; Domakonda and West, 2020). Previous investigations have demonstrated control of pyrimidine biosynthesis relative to enzyme synthesis and transcarbamoylase activity in taxonomically-related species (West, 2002; West, 2004a; West, 2009; Domakonda and West, 2020). This allows a direct comparison of regulation of pyrimidine biosynthesis in *P. chlororaphis* with the regulation of pyrimidine biosynthesis in taxonomically-related species. Another reason for investigating this microbe is that it has the potential to limit plant damage by fungal pathogens due to the compounds that it synthesizes. Similarly, this pseudomonad has also been shown to synthesize organic compounds that stimulate plant growth (Chin et al., 2005; Mehnaz et al., 2013; Mehnaz et al., 2014). Understanding an aspect of nucleic acid metabolism, such as pyrimidine biosynthesis, may help in genetically manipulating *P. chlororaphis* in order to isolate mutants that could block fungal pathogen growth while stimulating plant growth.

The results of this investigation indicated that exogenous orotic acid supplementation influenced the regulatory pattern of pyrimidine biosynthetic enzyme synthesis *P. chlororaphis* which contrasted the pattern of pyrimidine biosynthetic enzyme synthesis observed in the related taxonomically-related subspecies (Anzai et al., 2000; Mulet et al., 2010). It should be noted that succinate-grown *P. chlororaphis*, *P. taetrolens* or *P. aurantiaca* cells did exhibit some similarities to their patterns of enzyme synthesis when orotic acid was present in the medium (West, 2004a; Domakonda and West, 2020). With uracil addition to glucose minimal medium-grown *P. chlororaphis* cells, it was noticed that pattern of pyrimidine biosynthetic enzyme activities had some similarities to the pattern observed for *P. fragi* or *P. aurantiaca* (West, 2002). In uracil-supplemented, succinate-grown cells of *P. chlororaphis* ATCC 17414, the pattern of how its pyrimidine biosynthetic enzyme activities responded to uracil addition was largely the same as was noted relative to the pattern of enzyme activities in *P. taetrolens* or *P. aurantiaca* (West, 2004a; Domakonda and West, 2020). In contrast, the effect of supplementing uracil to succinate-grown *P. chlororaphis* cells on the pyrimidine biosynthetic enzyme activities was significantly different from the pattern observed in the related species *P. lundensis* (West, 2009). Based on the results from this study, it is obvious that *P. chlororaphis* has significant differences with other taxonomically-related species relative to orotic acid or uracil supplementation independent of carbon source (Anzai et al., 2000; Mulet et al., 2010). Overall, the effect of pyrimidine supplementation on the pyrimidine biosynthetic enzyme activities in *P. chlororaphis* was largely different than what has reported for the response of the pyrimidine biosynthetic enzyme activities in other members of its homology group relative to pyrimidine addition (West, 2002; West, 2004a; West, 2009; Domakonda and West, 2020).

Regulation of the pathway enzyme synthesis was previously investigated utilizing pyrimidine limitation experiments in the taxonomically-related species *P. lundensis*, *P. fragi* and *P. taetrolens* and *P. aurantiaca* (West, 2002; West, 2004a; West, 2009; Domakonda and West, 2020). Pyrimidine limitation experiments of pyrimidine auxotrophs need to be undertaken to decide if derepression of the pyrimidine biosynthetic enzyme activities can be seen (Lazzarini et al., 1969). Previous pyrimidine-limitation experiments were conducted with glucose or succinate as a carbon source to learn whether carbon source is a factor in the level of derepression of enzyme synthesis. In this study, a uracil auxotrophic strain BW-1, isolated from *P. chlororaphis* ATCC 17414, was found to be deficient for decarboxylase activity. It should be noted that there is a range of how long the pseudomonad cells can be starved for uracil during the pyrimidine limitation experiments. Considerable lysis of the *P. chlororaphis* strain BW-1 cells was observed after 2 hours of pyrimidine limitation which limited how long the mutant cells could be starved for uracil. In glucose-grown cells of *P. chlororaphis* strain BW-1 starved for uracil for 1 hour, pyrimidine limitation caused an increase in the four pathway enzymes activities by up to 8-fold. Interestingly, the glucose-grown cells of *P. chlororaphis* strain BW-1 limited for pyrimidine for 2 hours resulted in significant decreases in the 4 pathway enzyme activities by up to 83%. When the succinate-grown *P. chlororaphis* strain BW-1 cells were pyrimidine-limited for 1 or 2 hours of pyrimidine limitation, the four pyrimidine biosynthetic pathway pathway activities increased by up to 17-fold compared to the mutant cells grown in excess uracil. It appeared that a higher level of depression of enzyme synthesis was witnessed in the *P. chlororaphis* strain BW-1 cells limited for pyrimidines than what was previously noted following pyrimidine limitation of the mutant strains from *P. lundensis* and *P. taetrolens* (West, 2004a; West, 2009).

It has been reported that regulation of aspartate transcarbamoylase at the level of activity exists in species classified within the *P. chlororaphis* homology group (West, 2004a; West, 2009). This allowed a comparison of the regulation of aspartate transcarbamoylase activity in *P. chlororaphis* with other taxonomically-related species within the *P. chlororaphis* homology group (Anzai et al., 2000; Mulet et al., 2010). In this work, glucose-grown cell extracts of *P. chlororaphis* ATCC 17414 cells were prepared and analyzed as to how the activity of the transcarbamoylase in cell extracts was controlled. It was found that addition to the effectors pyrophosphate, uridine 5'-monophosphate, cytidine 5'-monophosphate, uridine 5'-diphosphate and guanosine 5'-triphosphate showed a high inhibitory effect on transcarbamoylase activity in *P. chlororaphis* ATCC 17414. Clearly, the finding of this study indicated that the regulation of *P. chlororaphis* ATCC 17414 aspartate transcarbamoylase activity was very different from how the *P. fragi*, or *P. taetrolens*, *P. lundensis* and *P. aurantiaca* aspartate transcarbamoylase activities were regulated (West, 2002; West, 2004a; West, 2009; Domakonda and West, 2020). There was greater inhibition of the transcarbamoylase in *P. fragi*, *P. taetrolens*, *P. lundensis* and *P. aurantiaca* by adenosine 5'-triphosphate and uridine 5'-triphosphate than was observed for the *P. chlororaphis* transcarbamoylase (West, 2002; West, 2004a; West, 2009; Domakonda and West, 2020).

Conclusion

It was concluded from the results collected in this study that the pattern of regulation of pyrimidine biosynthesis in *P. chlororaphis* ATCC 17414 differed from the patterns of regulation of pyrimidine biosynthesis exhibited in the taxonomically-related species *P. fragi*, *P. taetrolens*, *P. lundensis* and *P. aurantiaca* (Anzai et al., 2000; Mulet et al., 2010). Considering that the regulation of pyrimidine biosynthesis in *P. chlororaphis* was different from the the taxonomically-related species classified within the *P. chlororaphis* homology group, it may be necessary to further reclassify the species assigned within this homology group. With *P. chlororaphis* having possible commercial applications as a plant growth stimulant or as an agent that inhibits pathogenic fungal growth on plants, this study provides new information regarding the regulation of its nucleic acid metabolism which may provide useful in the genetic manipulation of its cells. Perhaps the modification of the nucleic acid metabolism in *P. chlororaphis* could allow the bacterium to inhibit plant pathogen growth while still stimulating plant growth. It is clear that the findings presented in this study could be of significant value as it pertains to the potential use of *P. chlororaphis* in agricultural applications.

Acknowledgment

This work was funded by the Welch Foundation Departmental Grant T-0014.

References

Adair, L. B., and Jones, M. E. (1972). Purification and characteristics of aspartate transcarbamylase from *Pseudomonas fluorescens*. *Journal of Biological Chemistry*, 247(8), 2308-2315.

Anzai, Y., Kim, H., Park, J. Y., Wakabayashi, H., and Oyaizu, H. (2000). Phylogenetic affiliation of the pseudomonads based on 16S rRNA sequence. *International Journal of Systematic and Evolutionary Microbiology*, 50(4), 1563-1589.

Beck, D. A., and O'Donovan, G. A. (2008). Pathways of pyrimidine salvage in *Pseudomonas* and former *Pseudomonas*: Detection of recycling enzymes using high-performance liquid chromatography. *Current Microbiology*, 56(2), 162-167.

Beckwith, J. R., Pardee, A. B., Austrian, R., and Jacob, F. (1962). Coordination of the synthesis of the enzymes in the pyrimidine pathway of *E. coli*. *Journal of Molecular Biology*, 5(6), 618-634.

Bradford, M. M. (1976). A rapid and sensitive method for the quantitation of microgram quantities of protein utilizing the principle of protein-dye binding. *Analytical Biochemistry*, 72(1-2), 248-254.

Chin, A. W. T. F., van den Broek, D., Lugtenberg, B. J., and Bloemberg, G. V. (2005). The *Pseudomonas chlororaphis* PCL1391 sigma regulator *psrA* represses the production of the antifungal metabolite phenazine-1-carboxamide. *Molecular Plant-Microbe Interactions*, 18(3), 244-253.

Chu, C. P., and West, T. P. (1990a). Pyrimidine biosynthetic pathway of *Pseudomonas fluorescens*. *Journal of General Microbiology*, 136(5), 875-880.

Chu, C. P., and West, T. P. (1990b). Pyrimidine ribonucleoside catabolism in *Pseudomonas fluorescens* biotype A. *Antonie van Leeuwenhoek, International Journal of General and Molecular Microbiology*, 57(4), 253-257.

Chunduru, J., and West, T. P. (2018). Pyrimidine nucleotide synthesis in the emerging pathogen *Pseudomonas monteilii*. *Canadian Journal of Microbiology*, 64(6), 432-438.

Condon, S., Collins, J. K., and O'Donovan, G. A. (1976). Regulation of arginine and pyrimidine biosynthesis in *Pseudomonas putida*. *Journal of General Microbiology*, 92(2), 375-383.

Dietrich, L. E., Price-Whelan, A., Petersen, A., Whiteley, M., and Newman, D. K. (2006). The phenazine pyocyanin is a terminal signalling factor in the quorum sensing network of *Pseudomonas aeruginosa*. *Molecular Microbiology*, 61(5), 1308-1321.

Domakonda, A., and West, T. P. (2020). Control of pyrimidine nucleotide formation in *Pseudomonas aurantiaca*. *Archives of Microbiology*, 202(6), 1551-1557.

Gill, R., and West, T. P. (2022). Control of a pyrimidine ribonucleotide salvage pathway in *Pseudomonas oleovorans*. *Archives of Microbiology*, 204(7), 383.

Haugaard, L. E., and West, T. P. (2002). Pyrimidine biosynthesis in *Pseudomonas oleovorans*. *Journal of Applied Microbiology*, 92(3) 517-525.

Kelln, R. A., Kinahan, J. J., Foltermann, K. F., and O'Donovan, G. A. (1975). Pyrimidine biosynthetic enzymes of *Salmonella typhimurium*, repressed specifically by growth in the presence of cytidine. *Journal of Bacteriology*, 124(2), 764-774.

Kim, S., and West, T. P. (1991). Pyrimidine catabolism in *Pseudomonas aeruginosa*. *FEMS Microbiology Letters*, 77(2-3), 175-179.

Lazzarini, R. A., Nakata, K., and Winslow, R. M. (1969). Coordinate control of ribonucleic acid synthesis during uracil deprivation. *Journal of Biological Chemistry*, 244(11), 3092-3100.

Morohoshi, T., Wang, W. Z., Suto, T., Saito, Y., Ito, S., Someya, N., and Ikeda, T. (2013). Phenazine antibiotic production and antifungal activity are regulated by multiple quorum-sensing systems in *Pseudomonas chlororaphis* subsp. aurantiaca StFRB508. *Journal of Bioscience and Bioengineering*, 116(5), 580-584.

Mehnaz, S., Baig, D. N., and Lazarovits, G. (2010). Genetic and phenotypic diversity of plant growth promoting rhizobacteria isolated from sugarcane plants growing in Pakistan. *Journal of Microbiology and Biotechnology*, 20(12), 1614-1623.

Mehnaz, S., Saleem, R. S., Yameen, B., Pianet, I., Schnakenburg, G., Pietraszkiewicz, H., Valeriote, F., Josten, M., Sahl, H. G., Franzblau, S. G., and Gross, H. (2013). Lahorenoic acids A-C, ortho-dialkyl-substituted aromatic acids from the biocontrol strain *Pseudomonas aurantiaca* PB-St2. *Journal of Natural Products*, 76(2), 135-141.

Mehnaz, S., Bauer, J. S., and Gross, H. (2014). Complete genome sequence of the sugar cane endophyte *Pseudomonas aurantiaca* PB-St2, a disease-suppressive bacterium with antifungal activity toward the plant pathogen *Colletotrichum falcatum*. *Genome Announcements*, 2(1), e01108-13.

Mulet, M., Lalucat, J., and Garcia-Vales, E. (2010). DNA sequence-based analysis of the *Pseudomonas* species. *Environmental Microbiology*, 12(6), 1513-1530.

Murahari, E. C., and West, T. P. (2019). The pyrimidine biosynthetic pathway and its regulation in *Pseudomonas jessenii*. *Antonie van Leeuwenhoek, International Journal of General and Molecular Microbiology*, 112(3), 461-469.

O'Donovan, G. A., and Neuhard, J. (1970). Pyrimidine metabolism in microorganisms. *Bacteriological Reviews*, 34(3), 278-343.

Pardee, A. B., and Yates, R. A. (1956a). Pyrimidine biosynthesis in *Escherichia coli*. *Journal of Biological Chemistry*, 221(2), 743-756.

Pardee, A. B., and Yates, R. A. (1956b). Control of pyrimidine biosynthesis in *Escherichia coli* by a feedback mechanism. *Journal of Biological Chemistry*, 221(2), 757-770.

Patten, C. L., and Glick, B. R. (2002). Role of *Pseudomonas putida* indoleacetic acid in development of the host plant root system. *Applied and Environmental Microbiology*, 68(8), 3795-3801.

Pohanka, A., Broberg, A., Johansson, M., Kenne, L., and Levenfors, J. (2005). Pseudotrienic acids A and B, two bioactive metabolites from *Pseudomonas* sp. MF381-IODS. *Journal of Natural Products*, 68(9), 1380-1385.

Prescott, L. M., and Jones, M. E. (1969). Modified methods for the determination of carbamyl aspartate. *Analytical Biochemistry*, 32(3), 408-419.

Price-Whelan, A., Dietrich, L. E., and Newman, D. K. (2006). Rethinking 'secondary' metabolism: physiological roles for phenazine antibiotics. *Nature Chemical Biology*, 2(2), 71-78.

Santiago, M. F., and West, T. P. (1999). Effect of nitrogen source on pyrimidine catabolism by *Pseudomonas fluorescens*. *Microbiological Research*, 154(3), 221-224.

Santiago, M. F., and West, T. P. (2002a). Regulation of pyrimidine synthesis in *Pseudomonas mendocina*. *Journal of Basic Microbiology*, 42(1), 75-79.

Santiago, M. F., and West, T. P. (2002b). Control of pyrimidine formation in *Pseudomonas putida* ATCC 17536. *Canadian Journal of Microbiology*, 48(12), 1076-1081.

Santiago, M. F., and West, T. P. (2003a). Comparison of aspartate transcarbamoylase regulation in *Pseudomonas alcaligenes* and *Pseudomonas mendocina*. *Journal of Basic Microbiology*, 43(1), 75-79.

Santiago, M. F., and West, T. P. (2003b). Influence of carbon source on pyrimidine synthesis in *Pseudomonas mendocina*. *Journal of Basic Microbiology*, 43(6), 534-538.

Santiago, M. F., and West, T. P. (2003c). Effect of carbon source on pyrimidine biosynthesis in *Pseudomonas alcaligenes* ATCC 14909. *Microbiological Research* 158(2), 195-199.

Schultheisz, H. L., Szymczyna, B. R., Scott, L. G., and Williamson, J. R. (2011). Enzymatic *de novo* pyrimidine nucleotide synthesis. *Journal of the American Chemical Society*, 133(2), 297-304.

Schwartz, M., and Neuhard, J. (1975). Control of expression of the *pyr* genes in *Salmonella typhimurium*: Effects of variations in uridine and cytidine nucleotide pools. *Journal of Bacteriology*, 121(3), 814-822.

Stanier, R. Y. (1947). Simultaneous adaptation: A new technique for the study of metabolic pathways. *Journal of Bacteriology*, 54(3), 339-348.

Watrin, L., Lucas, S., Purcarea, C., Legrain, C., and Prieur, D. (1999). Isolation and characterization of pyrimidine auxotrophs, and molecular cloning of the *pyrE* gene from the hyperthermophilic Archaeon *Pyrococcus abyssi*. *Molecular Genetics and Genomics*, 262(2), 378- 381.

Watson, J. M., and Holloway, B. W. (1976). Suppressor mutations in *Pseudomonas aeruginosa*. *Journal of Bacteriology*, 125(3), 780-786.

West, T. P. (1988). Metabolism of pyrimidine bases and nucleosides by *Pseudomonas fluorescens* biotype F. *Microbios*, 56(226), 27-36.

West, T. P. (1989). Isolation and characterization of thymidylate synthetase mutants of *Xanthomonas maltophilia*. *Archives of Microbiology*, 151(3), 220-222.

West, T. P. (1991). Pyrimidine base and ribonucleoside utilization by the *Pseudomonas alcaligenes* group. *Antonie van Leeuwenhoek, International Journal of General and Molecular Microbiology*, 59(4), 263-268.

West, T. P. (1992). Pyrimidine base and ribonucleoside catabolic enzyme activities of the *Pseudomonas diminuta* group. *FEMS Microbiology Letters*, 99(2-3), 305-310.

West, T. P. (1994a). Control of the pyrimidine biosynthetic pathway in *Pseudomonas pseudoalcaligenes*. *Archives of Microbiology*, 162(1-2), 75-79.

West, T. P. (1994b). Pyrimidine ribonucleoside catabolic enzyme activities of *Pseudomonas pickettii*. *Antonie van Leeuwenhoek, International Journal of General and Molecular Microbiology*, 66(4), 307-312.

West, T. P. (1996). Degradation of pyrimidine ribonucleosides by *Pseudomonas aeruginosa*. *Antonie van Leeuwenhoek, International Journal of General and Molecular Microbiology*, 69(4), 331-335.

West, T. P. (1997a). Pyrimidine biosynthesis in *Pseudomonas stutzeri* ATCC 17588. *Antonie van Leeuwenhoek, International Journal of General and Molecular Microbiology*, 72(3), 175-181.

West, T. P. (1997b). Regulation of aspartate transcarbamoylase activity in *Pseudomonas stutzeri*. *Microbiological Research*, 152(4), 373-375.

West, T. P. (2000). Role of cytosine deaminase and β-alanine-pyruvate transaminase in pyrimidine base catabolism by *Burkholderia cepacia*. *Antonie van Leeuwenhoek, International Journal of General and Molecular Microbiology*, 77(1), 1-5.

West, T. P. (2001). Pyrimidine base catabolism in *Pseudomonas putida* biotype B. *Antonie van Leeuwenhoek, International Journal of General and Molecular Microbiology*, 80(2), 163-167.

West, T. P. (2002). Control of pyrimidine synthesis in *Pseudomonas fragi*. *Letters in Applied Microbiology*, 35(5), 380-384.

West, T. P. (2004a). Regulation of pyrimidine nucleotide formation in *Pseudomonas taetrolens* ATCC 4683. *Microbiological Research*, 159(1), 29-33.

West, T. P. (2004b). Pyrimidine nucleotide synthesis in *Pseudomonas citronellolis*. *Canadian Journal of Microbiology*, 50(6), 455-459.

West, T. P. (2004c). Regulation of pyrimidine nucleotide formation in *Pseudomonas reptilivora*. *Letters in Applied Microbiology*, 38(2), 81-86.

West, T. P. (2005a). Regulation of the pyrimidine biosynthetic pathway in *Pseudomonas mucidolens*. *Antonie van Leeuwenhoek, International Journal of General and Molecular Microbiology*, 88(2), 181-186.

West, T. P. (2005b). Regulation of pyrimidine synthesis in *Pseudomonas resinovorans*. *Letters in Applied Microbiology*, 40(6), 473-478.

West, T. P. (2005c). Effect of carbon source on pyrimidine formation in *Pseudomonas fluorescens* ATCC 13525. *Microbiological Research*, 160(4), 337-342.

West, T. P. (2007a). Regulation of pyrimidine nucleotide biosynthesis in *Pseudomonas synxantha*. *Antonie van Leeuwenhoek, International Journal of General and Molecular Microbiology*, 92(3), 353-358.

West, T. P. (2007b). Regulation of pyrimidine formation in *Pseudomonas oryzihabitans*. *Journal of Basic Microbiology*, 47(5), 440-443.

West, T. P. (2009). Regulation of pyrimidine formation in *Pseudomonas lundensis*. *Canadian Journal of Microbiology*, 55(3), 261-268.

West, T. P. (2010a). Control of pyrimidine nucleotide formation in *Pseudomonas fulva*. *Antonie van Leeuwenhoek, International Journal of General and Molecular Microbiology*, Antonie van Leeuwenhoek 97(3), 307-311.

West, T. P. (2010b). Effect of carbon source on pyrimidine biosynthesis in *Pseudomonas oryzihabitans*. *Journal of Basic Microbiology*, 50(4), 397-400.

West, T. P. (2012). Pyrimidine biosynthesis in *Pseudomonas veronii* and its regulation by pyrimidines. *Microbiological Research*, 167(5), 306-310.

West, T. P. (2014). Pyrimidine nucleotide synthesis in *Pseudomonas nitroreducens* and the regulatory role of pyrimidines. *Microbiological Research*, 169(12), 954-958.

West, T. P., and O'Donovan, G. A. (1982). Repression of cytidine triphosphate synthetase in *Salmonella typhimurium* by pyrimidines during uridine nucleotide depletion. *Journal of General Microbiology*, 128(4), 895-899.

West, T. P., Herlick, S. A., and O'Donovan, G. A. (1983). Inverse relationship between thymidylate synthetase and cytidine triphosphate synthetase activities during pyrimidine limitation in *Salmonella typhimurium*. *FEMS Microbiology Letters*, 18(3), 275-278.

Xu, G., and West, T. P. (1992). Reductive catabolism of pyrimidine bases by *Pseudomonas stutzeri*. *Journal of General Microbiology*, 138(11), 2459-2463.

Chapter 4

Pyrimidines as Potential Corrosion Inhibitors: Recent Developments and Future Perspectives

Dheeraj Singh Chauhan[1,2,*]
M. A. Quraishi[3]
Chandrabhan Verma[3]
and V. S. Saji[3]

[1]Modern National Chemicals, Second Industrial City, Dammam, Saudi Arabia
[2]Center of Research Excellence in Corrosion, Research Institute,
King Fahd University of Petroleum and Minerals, Dhahran, Saudi Arabia
[3]Interdisciplinary Research Center for Advanced Materials,
King Fahd University of Petroleum and Minerals, Dhahran, Saudi Arabia

Abstract

Heterocyclic molecules containing nitrogen are extensively studied as corrosion inhibitors due to their capability to form strong coordination bonds with metallic atoms. The excellent corrosion protection action of pyrimidine derivatives can be attributed to the presence of electron-withdrawing and electron-donating substituents that allow an improved protection performance compared to that of the parent pyrimidine molecule. Computational studies have provided evidence that the pyrimidine-based molecules undergo donor-acceptor type interactions with metallic substrates. In general, the organic corrosion inhibitors undergo a mixed-type of adsorption behavior, and their adsorption follows the Langmuir isotherm. This chapter presents an overview of

---

[*] Corresponding Author's Email: dheeraj.chauhan.rs.apc@itbhu.ac.in

In: Pyrimidines and Their Importance
Editor: Roger G. Ward
ISBN: 979-8-88697-656-4
© 2023 Nova Science Publishers, Inc.

pyrimidine-based molecules as corrosion inhibitors for metals and alloys in a variety of corrosive environments.

Keywords: diazines, pyrimidines, heterocycles, corrosion inhibitor, adsorption

Introduction

Metals and alloys find wide applicability in a wide range of industries. In several of these, the surfaces of these metallic materials come in direct contact with extreme environments, viz., concentrated acids, alkali, saline media, etc. (Quraishi, Chauhan, and Saji 2021, Quraishi, Chauhan, and Saji 2020). This leads to deteriorating attacks on these metal surfaces that can cause huge economic losses and potential structural failure. The major strategies used to counter this situation involve the use of corrosion-resistant alloys, anti-corrosion coatings, cathodic protection, and corrosion inhibitors (Cottis 2010, Revie 2011). Corrosion inhibitors in aqueous corrosive environments adsorb on the metal surfaces and form a protective film that influences retardation in the corrosion rate. Various kinds of inorganic and organic molecules are reported as effective inhibitors in a wide range of corrosive media.

Conventionally, for the protection of metals and alloys against corrosion, the organic corrosion inhibitors that are used come from the classes of the heterocyclic molecules bearing the N, S, O etc. heteroatoms in the organic rings, and hence called as heterocycles (Joule and Mills 2012, Eicher, Hauptmann, and Speicher 2013). The heterocycles exist in a large variety of biomolecules, e.g., pharmaceutical products (Singh et al. 2017, Dohare, Chauhan, and Quraishi 2018), vitamins, amino acids, natural extracts (Chaubey et al. 2020), biological polymers, etc. (Ansari et al. 2020, EL Mouaden et al. 2020, El Mouaden et al. 2018, Mouaden et al. 2020). A number of these heterocycles, especially from the categories of the five and six-membered ring-based compounds, have been utilized as corrosion inhibitors (Quraishi, Chauhan, and Saji 2020). The major categories of heterocyclic corrosion inhibitors are imidazoles, triazoles, oxadiazoles, thiadiazoles, pyridines, pyrimidines, and their benzene fused molecules such as the benzimidazole, benzotriazole, benzothiazole, etc. (Onyeachu et al. 2020, ElBelghiti et al. 2016, Xu et al. 2014, Anusuya et al. 2017, Kaya et al. 2016).

Pyrimidine is a heterocyclic aromatic organic compound containing two nitrogen atoms at positions 1 and 3 of a six-membered ring (Katritzky et al. 2010). It is similar in structure to pyridine, which contains one N atom and belongs to the family of diazines having the other two members as pyrazine and pyridazine, each differing in the position of the N atoms in the aromatic ring. The pyrimidine-based organic compounds have been significantly explored as inhibitors for metallic corrosion in a wide variety of acidic as well as neutral environments. In the present chapter, the biological significance of pyrimidines and their application in corrosion inhibition is discussed. A review of the literature is presented on the pyrimidine derivatives on the corrosion inhibition behavior in the acidic and the neutral media on different metallic materials. Some of the computational insights on the adsorption and protection behavior of pyrimidines in various corrosive environments is discussed herein.

An Overview of Corrosion and the Significance of Heterocyclic Corrosion Inhibitors

Corrosion can be explained as the degradation or deterioration of a metallic material when it comes in contact with a corrosive environment. The metal-based materials are produced after the purification of metals from their respective ore forms, such as oxides, sulfides, etc., and are employed in developing various alloys and other metallic structures. Therefore, these structures show a strong tendency to undergo corrosion and revert to their naturally occurring ore form (Sastri 1998, Priyanka et al. 2022, Joshi et al. 2022). This process is aggravated in concentrated aqueous environments such as strong acids, alkali, or saline media. Hence, the process of corrosion can also be described as: '*metallurgy in reverse.*' The major factors influencing the corrosion of metals include the concentration of the corrosive media, temperature, flow rate, surface composition of the metal and alloy, etc. On a global scale, Corrosion causes a major economic impact amounting to 276 billion US$ annually, which comes to almost 3.4% of the world's GDP (Quraishi, Chauhan, and Saji 2020). Several methods are in practice to control and mitigate corrosion, including the use of (i) corrosion-resistant alloys, (ii) anti-corrosion coatings, and (iii) corrosion inhibitors.

A corrosion inhibitor can be defined as a chemical additive that, when introduced to a corrosive environment, considerably brings down the rate of aqueous corrosion, without modifying the concentration and pH of the environment (Quraishi and Chauhan 2022b, a, Quraishi, Chauhan, and Ansari 2021). Due to the ease of application, several organic molecules are in use as corrosion inhibitors. These molecules function by adsorbing on the target metallic surface and forming a thin protective film that significantly lowers the rate of corrosion by restricting the access of the corrosive electrolyte to the metal surface. The salient features of a good organic corrosion inhibitor are the presence of an abundant amount of heteroatoms (N, S, O), π-bonds, conjugation, aromatic rings, etc. The organic heterocyclic molecules are, therefore, considered effective corrosion inhibitors (Chauhan, Mouaden, et al. 2020, Xiong et al. 2019, Chauhan, Sorour, and Quraishi 2016, Singh et al. 2019a). Here an overview of the pyrimidine-based corrosion inhibitors for various corrosive environments is presented.

Pyrimidine and Its Derivatives as Corrosion Inhibitors

Pyrimidine derivatives come under the category of six-membered heterocyclic systems. By replacing two CH units in the benzene ring, the three possible diazine structures namely pyridazine (1,2-diazine), pyrimidine (1,3-diazine), and pyrazine (1,4-diazine) can be obtained (Eicher, Hauptmann, and Speicher 2013, Katritzky et al. 2010). Diazines (Figure 1) are aromatic six-membered heterocycles that contain two sp^2-hybridized nitrogen atoms in the ring. Due to the presence of heteroatom N in the ring, these molecules come under the category of heterocyclic systems and are fully aromatic.

Compared to pyridines that contain one N atom in the ring, the presence of two N atoms allows the pyrimidines an additional site for interaction with the metallic surface. The six-membered heterocyclic ring of pyrimidine can be considered one of the most stable heterocyclic structures owing to the presence of two N atoms, each having a lone pair of electrons. The organic corrosion inhibitors generally act by donating the lone pair electrons with the unoccupied d-orbitals of the target metal atoms, leading to the formation of coordinate bonds.

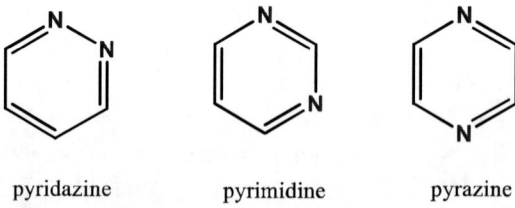

pyridazine   pyrimidine   pyrazine

Figure 1. Molecular structures of the three diazines.

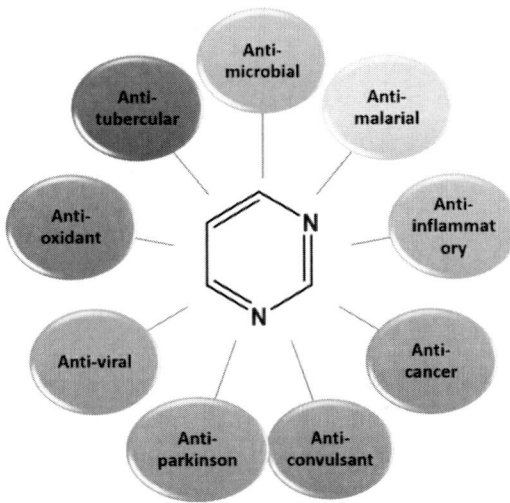

Figure 2. Medicinal applications of pyrimidines.

Alternatively, the heterocycles can also accept the free electrons from the metal surface by involving their anti-bonding orbitals to form feedback bonds. Therefore, the pyrimidine ring is expected to perform excellent adsorption and inhibition at the metal/solution interface. Furthermore, the pyrimidine ring is a well-known constituent of the nucleobases cytosine (C), thymine (T), and uracil (U), constituting the DNA. Pyrimidine pharmacophore is an important and integral part of DNA and RNA and play an essential role in several biological processes; and have considerable chemical and pharmacological utility as antibiotics, antibacterial, cardiovascular as well as agrochemical, and veterinary product (Reznik et al. 2008, Kumar et al. 2009, Trivedi et al. 2010, Bazgir, Khanaposhtani, and Soorki 2008, Lagoja 2005). The pyrimidine ring is also found in the Vitamins such as thiamine, folic acid, and riboflavin. The wide variety of biological activities observed for these compounds turned pyrimidine derivatives into environmentally benign compounds. Figure 2 shows the medicinal applications of pyrimidine derivatives.

Synthesis of Some Pyrimidine-Based Corrosion Inhibitors

The synthesis of organic corrosion inhibitors, in general, is accomplished using multiple-step synthesis that requires expert scientific personnel. However, considering the large-scale application, as in the case of corrosion inhibitors, it is necessary to maintain cost-effectiveness. Further, the synthesis of the inhibitors should be in accordance with the Green Chemistry principles and follow the environmental regulations. Herein, the multicomponent reactions (MCRs) represent the method in which, instead of a multiple-step synthesis, the synthesis is performed involving three or more reactants in a single step, leading directly to the final product (Alvim, da Silva Junior, and Neto 2014, Verma et al. 2018).

Modern technology makes use of microwave (MW) (Tierney and Lidström 2009, Kappe 2008, 2004) and ultrasound (US)-based organic synthesis techniques for developing organic corrosion inhibitors (Kimura 2015, Singh, Shukla, and Quraishi 2011). MW method works at 2.45 GHz involving the ionic conduction and dipolar polarization phenomena (Kappe 2004, Polshettiwar and Varma 2008). This method avoids the generation of localized thermal gradients and allows uniform heating of the reaction mixture.

Synthetic scheme for the pyrimidine derivatives (Verma et al. 2017).

Preparation of 3,4-dihydropyrimidin-2(1H)-ones by Biginelli reaction (Yadav, Maiti, and Quraishi 2010).

Synthetic route of 5-ethyl 4-(4-methoxyphenyl)-6-methyl-2-thioxo-1, 2, 3, 4 tetrahydropyrimidine-5-carboxylate (Korde et al. 2015).

Scheme for the synthesis of glucosamine-based, pyrimidine-fused heterocycles (Verma, Olasunkanmi, Ebenso, et al. 2016).

Synthetic scheme of hexahydropyrimido-pyrimidinones (Haque et al. 2017).

On the other hand, the US method involves an ultrasonic probe that oscillates at a high frequency, resulting in frequent collisions and localized temperature variations in the reaction vessel, thereby improving the reaction speed (Kimura 2015). MW and US are considered excellent alternatives for conventional reflux-based organic synthesis. The usage of these techniques fulfills the provisos of Green technology (Polshettiwar and Varma 2008, Chauhan, Verma, et al. 2020). Several MCR-based syntheses of organic compounds to be used as corrosion inhibitors are reported in the literature. Herein, we have shown some of the representative schemes for the synthesis of some pyrimidine derivatives as corrosion inhibitors.

Scheme for the synthesis of pyrimidopyrimidinones (Ansari et al. 2015).

Synthetic route of 5-arylpyrimido-[4,5-b]quinoline-diones (Verma, Olasunkanmi, Obot, et al. 2016).

Literature Survey and Case Studies

A thorough understanding of the corrosion inhibition action of an organic molecule can be analyzed with a combination of various techniques. These include the weight loss tests (gravimetric measurements) wherein a known weight metal sample is immersed in the corrosive solution in the absence and the presence of the inhibitor for a fixed period of time at a given temperature (Baboian 2005). The difference in the weights before and after immersion allows the determination of the corrosion rates. These measurements at various temperatures and for varying inhibitor concentrations allow the measurements of kinetic and thermodynamic parameters relevant to the inhibitor adsorption. The next group includes the electrochemical measurements using the techniques of electrochemical impedance spectroscopy (EIS) (Dhillon and Kant 2017, Jorcin et al. 2006), electrochemical frequency modulation (EFM) (Obot and Onyeachu 2018, Khaled 2008), linear polarization resistance (LPR), and potentiodynamic polarization (PDP) (Frankel and Rohwerder 2007, McCafferty 2005). These techniques shed further light on the inhibitor adsorption mechanism and can also provide information on the corrosion rates. The third group of techniques includes surface analytical techniques, viz. water contact angle (WCA) measurements, scanning electron microscopy (SEM), atomic force microscopy (AFM), x-ray photoelectron spectroscopy (XPS), x-ray diffraction (XRD), etc. These techniques can provide information on the hydrophobicity of the metallic substrate prior to and after inhibitor adsorption (WCA); can visualize the surface of the metal to show the improvement in the surface smoothness after inhibitor adsorption (SEM); and the related quantitative surface roughness, with 3-d surface topography (AFM); corrosion products and the metal-inhibitor complex formation (XRD, XPS).

The mild steel/1M HCl system has been the most commonly studied for corrosion inhibition studies. Three D-glucose derivatives of dihydropyrido [2,3-d:6,5-d′] dipyrimidine-2, 4, 6, 8(1H,3H,5H,7H)-tetraones were studied for mild steel in 1M HCl solution employing experimental and computational studies (Verma et al. 2017). The inhibitors GPH-1, GPH-2, GPH-3 contained –H, –OH, and –OCH$_3$ substituents. The inhibitor GPH-3, due to the presence of the electron-donating substituent, resulted in the highest inhibition efficiency at a modest inhibitor dose. The inhibition effect of 3,4-dihydropyrimidin-2(1H)-ones (DHPMs) was evaluated on the surface of mild steel in 1M HCl solution using gravimetric, electrochemical, surface, and computational studies (Yadav, Maiti, and Quraishi 2010). All three studied inhibitors showed high efficiencies at a quite low dose of 10 mgL^{-1}. The presence of $-NO_2$ substituent at *meta* and *para* positions lowered the protection performance, whereas the unsubstituted derivative, i.e., DHPM-3, produced a high performance. The synergistic corrosion inhibition performance of dihydropyrimdinones was evaluated for carbon steel in 0.5M H_2SO_4 using gravimetric as well as electrochemical and surface studies (Singh et al. 2019a). The presence of the $-OCH_3$ group resulted in a better performance, which improved further upon the introduction of iodide ions as synergistic agents. The influence of variation in the test temperature was carried out using PDP study. The obtained results shown in Figure 3 reveal that the introduction of the DHPMs to the corrosive solution does not cause a significant shift in the position of the E_{corr} in the case of DHPM-1 and DHPM-2, indicating a mixed-type behavior. On the other hand, for DHPM-3, a clear-cut shift in E_{corr} is observed, suggesting a purely cathodic mechanism. High efficiency of 98.15% was achieved with the inhibitor DHPM-3 in the presence of 3.01 ×10^{-4} molL^{-1} KI. 5-ethyl-4-(4-methoxyphenyl)-6-methyl-2-thioxo-1, 2, 3, 4-tetrahydropyrimidine-5-carboxylate was evaluated as a corrosion inhibitor for aluminium in 1M HCl solution and 83.33% protection efficiency was achieved at 500 mgL^{-1} (Korde et al. 2015). SEM and EDX analyses supported the inhibitor adsorption and the formation of a protective film on the metallic substrate. A series of four glucosamine-based, pyrimidine-fused heterocycles was studied on mild steel in 1M HCl environment employing experimental and computational studies (Verma, Olasunkanmi, Ebenso, et al. 2016). The authors studied the influence of –OH, -CH$_3$, and –NO$_2$ substituent groups on inhibition efficiency. The presence of –OH group leads to the highest inhibition efficiency. All the studied inhibitors showed high solubility in the corrosive environment, and the maximum efficiency was obtained at a very low inhibitor dosage. Two hexahydropyrimidopyrimidine derivatives were studied for corrosion protection performance on N80 steel in 15% HCl solution employing gravimetric measurements and electrochemical studies (Haque et al. 2017). Further, SEM and AFM studies were conducted, followed by computational studies, which supported an improved protection performance of the derivative containing the cinnamaldehyde moiety compared to that containing the hydroxyphenyl moiety.

A series of thioxo-pyrimido-pyrimidine derivatives (PPD) were also evaluated on mild steel in 1M HCl solution (Ansari et al. 2015). In addition to the variation in the substituent group, the authors also studied the effect of O vs S atom in the molecule. It was observed that the presence of –OCH$_3$ group was the most conducive to the protective performance, whereas the existence of S instead of O resulted in better performance.

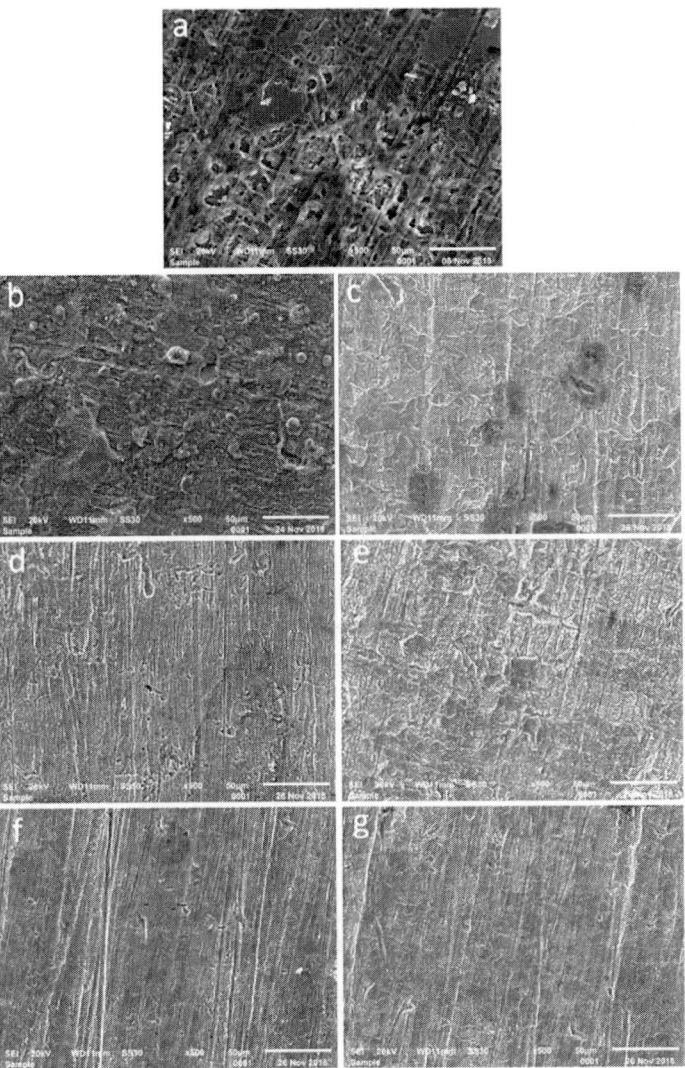

Figure 3. SEM images for carbon steel sample after immersion in (a) 0.5 M H_2SO_4 and (b, c) DHPM-1 and DHPM-1 + 3.01 × 10^{-4} M KI, (d, e) DHPM-2 and DHPM-2 + 3.01 × 10^{-4} M KI, (f, g) DHPM-3 and DHPM-3 + 3.01 × 10^{-4} M KI. Reproduced from Ref. (Singh et al. 2019a) 2019 with permission from Elsevier.

Table 1. Corrosion protection performance of some pyrimidine derivatives

System	Metal surface/ Electrolyte	(Conc.)/IE%	Adsorption behavior	Reference
D-glucose derivatives of dihydropyrido-dipyrimidine-tetraone GPH-1 GPH-2 GPH-3	Mild steel/ 1M HCl	$(10.15 \times 10^{-5}$ $molL^{-1})$ 93.91 95.21 97.82	Langmuir isotherm/ mixed type	(Verma et al. 2017)
Dihydropyrimidones DHPM-1 DHPM-2 DHPM-3	Mild steel/ 1M HCl	$(10$ $mgL^{-1})$ 94.2 92.0 98.8	Langmuir isotherm/ mixed type	(Yadav, Maiti, and Quraishi 2010)
Dihydropyrimdinones DHPM-1 DHPM-2 DHPM-3 DHPM-1 + KI DHPM-2 + KI DHPM-3 + KI	Mild steel/ 0.5M H_2SO_4	$(3.27 \times 10^{-5}$ $molL^{-1})$ 89.84 94.91 97.52 $(+ 3.01 \times 10^{-4}$ $molL^{-1}$ KI) 93.51 97.33 98.15	Langmuir isotherm/ mixed type	(Singh et al. 2019a)
5-ethyl 4-(4-methoxyphenyl)-6-methyl-2-thioxo-tetrahydropyrimidine-carboxylate	Aluminium/ 1M HCl	$(500\ mgL^{-1})$ 83.3%	Langmuir isotherm/ mixed type	(Korde et al. 2015)
Pyrimidine fused heterocycles CARB-1 CARB-2 CARB-3 CARB-4	Mild steel/ 1M HCl	$(7.41 \times 10^{-5}$ $molL^{-1})$ 88.69 90.86 93.47 96.52	Langmuir isotherm/ mixed type	(Verma, Olasunkanmi, Ebenso, et al. 2016)
Hexahydropyrimido[4,5-d]-pyrimidin-4(1H) ones PP-1 PP-2	N80 steel/ 15% HCl	$(250\ mgL^{-1})$ 88.3 77.2	Langmuir isotherm/ mixed type	(Haque et al. 2017)
Phenyl-pentahydro-pyrimidopyrimidine-trione PPD-1 PPD-2 PPD-3 PPD-4	Mild steel/ 1M HCl	$(400\ mgL^{-1})$ 97.1 95.7 93.3 88.0	Langmuir isotherm/ mixed type	(Ansari et al. 2015)

System	Metal surface/ Electrolyte	(Conc.)/IE%	Adsorption behavior	Reference
Thiopyrimidine derivatives TP-1 TP-2 TP-3 TP-4	Mild steel/ 1M HCl	(250 mgL^{-1}) 91.90 95.70 96.66 98.57	Langmuir isotherm/ mixed type	(Singh, Singh, and Quraishi 2016)
Arylpyrimido-[4,5-b]quinoline-diones APQD-1 APQD-2 APQD-3 APQD-4	Mild steel/ 1M HCl	(20 mgL^{-1}) 92.17 93.91 95.65 96.52	Langmuir isotherm/ mixed type	(Verma, Olasunkanmi, Obot, et al. 2016)
Pyranopyrimidine derivatives (Condensed Uracils) CU-1 CU-2 CU-3 CU-4	Mild steel/ 1M HCl	(350 mgL^{-1}) 78.0 66.1 95.5 85.2	Langmuir isotherm/ mixed type	(Yadav and Quraishi 2012)

All the studied four molecules showed a mixed-type behavior with cathodic predominance. Thiopyrimidines derivatives viz. 5-cyano-6-phenyl-2-thioxo-2,3-dihydropyrimidin-4-one (TP-1), 5-cyano-2-thioxo-6-(p-tolyl)-2,3 dihydropyrimidin-4-one (TP-2), 5-cyano-6-(4-methoxyphenyl)-2-thioxo-2,3-dihydropyrimidin-4-one (TP-3) 6-(4-(dimethylamino)phenyl)-5-cyano-2-thioxo-2,3-dihydropyrimidin-4-one (TP-4) were also evaluated for mild steel in 1M HCl solution (Singh, Singh, and Quraishi 2016). The presence of dimethylamino group led to a better inhibition performance with 98.57% protection at a dose of 250 mgL^{-1}. Corresponding SEM images of the blank and the inhibitor-adsorbed steel sample show an improvement in surface homogeneity after inhibitor adsorption, suggesting the formation of an inhibitor film. The XPS studies supported the formation of Fe–O bond formation in the case of FeO, Fe_2O_3 (Fe^{3+} oxide), Fe_3O_4, FeOOH (oxyhydroxide), and $FeCl_3$ as the corrosion products. C–N, C=N, and C–O bonding was also revealed to support the adsorbed inhibitors on the steel surface. Four 5-Arylpyrimido-[4,5-b]quinoline-diones were evaluated as corrosion inhibitors for mild steel in 1M HCl solution using experimental and computational studies (Verma, Olasunkanmi, Obot, et al. 2016). It was shown that the presence of –OH revealed a superior inhibition performance. Furthermore, it was observed that the presence of two –OH groups led to better performance. A series of four condensed uracils (CU) were also examined on mild steel in 1M HCl solution (Yadav and Quraishi 2012). The presence of –OCH$_3$ group in the molecular framework resulted in the highest inhibition efficiency. The increase in the molecule's electron density due to the electron-donating effect of the –OCH$_3$ group was attributed to this observation. All these molecules were composed of a pyran ring, a pyrimidine ring and a phenyl ring differing in the presence of substituent groups. UV-Vis studies for these inhibitors revealed the formation of metal-inhibitor complex. Table 1 shows a collection of names, the nature of metal/electrolyte, and the corrosion inhibition performance of some pyrimidine derivatives.

Theoretical Insight on Corrosion Inhibition Property of Pyrimidine-Based Compounds

Computational studies and modelings of corrosion inhibition effects of organic compounds have gained particular attention in recent years. DFT-based studies have elucidated several reactivity parameters such as the energies of the frontier molecular orbitals (FMO), E_{HOMO} (energy of the highest occupied molecular orbital), E_{LUMO} (energy of the lowest unoccupied molecular orbital), molecular orbital energy gap ($\Delta E = E_{LUMO} - E_{HOMO}$), hardness ($\eta$), electronegativity ($\chi$), softness ($\sigma$), dipole moment ($\mu$), nucleophilicity ($\omega$), and the fraction of electron transfer (ΔN), etc. (Dohare, Chauhan, Hammouti, et al. 2017, Dohare, Chauhan, Sorour, et al. 2017, Singh et al. 2019b, Obot, Macdonald, and Gasem 2015). The detailed description, significance, and calculation of these parameters are provided elsewhere. Although, it can be mentioned here that the increase in E_{HOMO} can be connected with the tendency to donate electrons from the inhibitor to the metal surface. On the other hand, a lowering in the E_{LUMO} value can be linked with the inclination towards electron acceptance by the inhibitor. The decrease in the ΔE can be directly correlated to the increase in the reactivity of a given inhibitor molecule, which in other words, can be reflected in terms of superior adsorption. Higher values of softness and nucleophilicity can be correlated to the high corrosion inhibition effectiveness and vice versa. On the other hand, lower electronegativity and hardness are consistent with greater inhibition effectiveness. Following the DFT analysis, the key sites related to inhibitor adsorption can be identified in the form of the FMO. A number of literature reports are available describing the inhibition effect of pyrimidine-based corrosion inhibitors using the DFT technique.

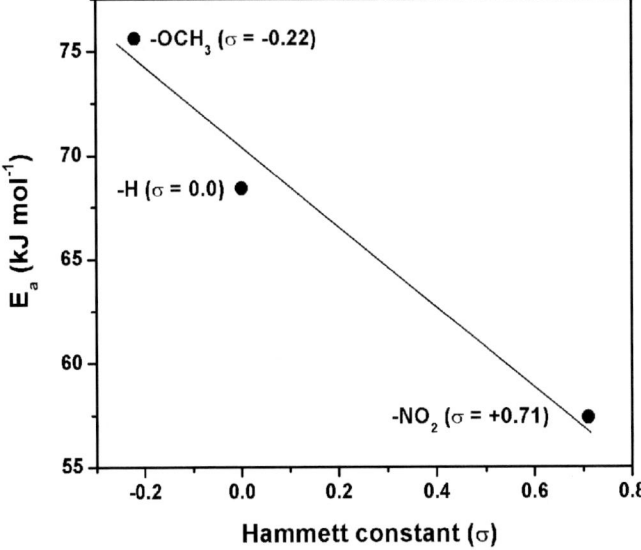

Figure 4. Correlation between Hammett constant and energy of activation for the studied pyrimidine derivatives (DHPM) (Singh et al. 2019a). Reproduced from Ref. (Singh et al. 2019a) 2019 with permission from Elsevier.

Molecular simulation techniques, namely molecular dynamics (MD) and Monte Carlo (MC) simulations are extensively used in testing and inhibiting the behavior of organic molecules (Obot, Haruna, and Saleh 2019). These techniques allow the study of the orientation of the corrosion inhibitor on the metallic substrate and provide the various energy parameters related to the metal-inhibitor interaction. A big advantage of these techniques is that the simulation of the adsorption of the inhibitor can be carried out at desired temperature and electrolyte composition. One of the most important parameters is the Adsorption energy (E_{ads}). A negative value of the E_{ads} is consistent with spontaneous adsorption and vice versa. In general, the inhibitor protection efficiency rises with an increase in the E_{ads} values. Literature study shows that MD and MC simulations have been considerably employed in studying the inhibition performance of pyrimidine derivatives.

For the studied inhibitors of GPH series, the obtained values of E_{HOMO}, electronegativity (χ), and dipole moment (μ) followed the experimental order of inhibition efficiency (Verma et al. 2017). The MC study revealed that the E_{ads} values rose upon inhibitor adsorption in the case of –OH, and –OCH$_3$ inhibitors supporting the electron-donating influence. For the studied series of

DHPMs on mild steel in 1M HCl, DFT analysis was carried out, which revealed that the inhibitor DHPM-3, devoid of the electron-withdrawing $-NO_2$ had the smallest value of the ΔE (Yadav, Maiti, and Quraishi 2010). This supported the experimental findings and corroborated the highest inhibition performance. For the studied DHPMs on mild steel in 0.5M H_2SO_4, it was observed that the ΔE values decreased with protonated forms of the inhibitors compared to that in the neutral form (Singh et al. 2019a). This indicated a greater reactivity in the protonated form for the inhibitors and supported the synergistic interaction with the iodide anions, leading to the adsorption and protection effect enhancement. Furthermore, the obtained trend with the energy of activation (E_a) values was in accordance with that of the Hammett substitution constants for the studied inhibitors (Figure 4). MC simulations supported that the DHPM-3 exhibited a better adsorption behavior compared to the other two inhibitors and acted mainly in the protonated form (Figure 5).

In the case of glucosamine-based, pyrimidine-fused heterocycles, the quantum chemical calculations provided evidence for the donor-acceptor type interaction between metal and the inhibitor molecules (Verma, Olasunkanmi, Ebenso, et al. 2016). MC studies revealed that the inhibitors underwent a flat orientation on the metallic substrate, and the predicted energy parameters followed that of the observed corrosion inhibition efficiency. In the case of the pyrimidine derivatives varying in the presence of cinnamaldehyde and benzaldehyde moieties, for the protonated inhibitor forms, the values of ΔE lowered, indicating a likelihood of inhibitor protonation due to the acid solution (Haque et al. 2017). The electrostatic potential (ESP) maps of the inhibitors in the neutral and protonation forms revealed a tendency of electron donation in the neutral form while a tendency of electron acceptance in the protonated forms. The MC study revealed that the cinnamaldehyde derivative assumed a parallel orientation to the steel surface (leading to a greater surface coverage), whereas the benzaldehyde derivative aligned in a manner perpendicular to the metal surface (leading to a lower surface coverage), thus supporting the trends observed in the experimental studies. The DFT estimations on the thiopyrimidine derivatives (TP) revealed a considerable lowering in the ΔE values with an increase in the extent of electron donation among the studied corrosion inhibitors (Singh, Singh, and Quraishi 2016). This trend was also in direct accordance with the rise in the E_{HOMO} values of the four studied inhibitors. DFT studies revealed in the case of APQDs, the electron donation trends prevailed from the inhibitor to the metal surface (Verma, Olasunkanmi, Obot, et al. 2016). The Fukui indices revealed that the prospective sites for the electron donation were the heteroatoms (N and O) for the studied molecules. The MD simulations indicated a parallel orientation of the four studied inhibitors on the metal surface resulting in high adsorption energies, which were in accordance with the trend determined experimentally. High negative values of the Mulliken charges observed on N and O atoms in the DFT studies for the CU molecules indicated their likelihood as the significant adsorption sites (Yadav and Quraishi 2012). This also supported the electron donation from the inhibitor molecules to the metal surface. CU-3 had the highest dipole moment among the studied molecules, which also supported its strong adsorption and greater inhibition on the mild steel surface.

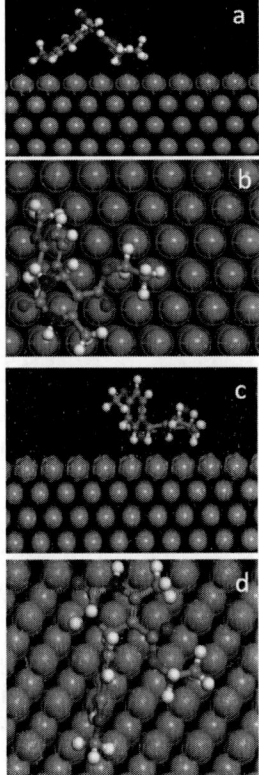

Figure 5. Side (a, c) and top (b, d) views of the most stable low-energy configurations for the adsorption of neutral and protonated forms of DHPM-3 on the Fe (110) surface (Singh et al. 2019a). Reproduced from Ref. (Singh et al. 2019a) 2019 with permission from Elsevier.

Challenges and Limitations

Although pyrimidine and its derivatives are extensively used as corrosion inhibitors for different metals and alloys in various electrolytes, however, most such compounds are toxic in nature. Therefore, considering the growing interest in green and sustainable development, the synthesis and application of environmentally sustainable derivatives should be preferred. A major associated disadvantage of pyrimidine and its derivatives preparation is that conventionally these are synthesized employing multiple-step reactions using toxic chemicals and solvents. This type of synthetic protocol are not preferable due to its association with various non-beneficial aspects such as low yield, tedious and time-consuming procedures, and environmentally hazardous properties. It is also noteworthy to mention that pyrimidine itself is not very much useful as a corrosion inhibitor. However, its derivatives prepared by the introduction of π-bonds, additional heteroatoms, heterocycles, phenyl rings, varying carbon chains, etc., are reported to yield excellent corrosion inhibitors. The inhibition performance of these molecules can be further enhanced through the introduction of external additives such as KI.

Summary and Outlook

Several research articles have been reported in the literature on the application of pyrimidine-based molecules as corrosion inhibitors for various metals and alloys. The presence of two N atoms in the pyrimidine rings aids in the adsorption. The presence of other heterocycle rings, additional phenyl rings, hydrophobic alkyl chains, and substituent groups allow an efficient coverage of the metallic substrate. Herein we have provided an overview of the various structural analogs of pyrimidines in various medicinal applications. Various strategies for the synthesis of pyrimidine derivatives have been elucidated focusing mostly on the MCR pathways to allow the readers to be acquainted with some of the simple strategies for the preparation of pyrimidine-based organic molecules. There is a further need to put emphasis on the green chemistry metrics with the estimation of toxicity, biodegradation, and bioaccumulation tests using animal models to obtain complete information on the pyrimidine-based corrosion inhibitors.

It is evident that most of the literature reports focus on the pyrimidine derivatives as corrosion inhibitors that target the metals and alloys for acidic environments. Further, among the acid solutions, most studies have been conducted on steel alloys in HCl solutions. These studies mostly aim to simulate acid-cleaning conditions. The available literature is scarce on the application of pyrimidine-based corrosion inhibitors for strongly acidic environments, such as 15-28% HCl, 20% H_2SO_4, which can model the industrial oil-well acid stimulation process. Some of the prospective application areas for the pyrimidine-based corrosion inhibitors could be in strongly saline environments, in the sweet corrosion of steel, and in alkaline environments. In addition, more studies on synergistic corrosion inhibition also should be reported on the pyrimidine derivatives.

Disclaimer

None.

References

Alvim, Haline GO, Eufranio N da Silva Junior, and Brenno AD Neto. "What Do We Know About Multicomponent Reactions? Mechanisms and Trends for the Biginelli, Hantzsch, Mannich, Passerini and Ugi Mcrs." *RSC Advances* 4, no. 97 (2014): 54282-99.

Ansari, Kashif Rahmani, Dheeraj Singh Chauhan, Mumtaz Ahmad Quraishi, Mohammad Abu Jafar Mazumder, and Ambrish Singh. "Chitosan Schiff Base: An Environmentally Benign Biological Macromolecule as a New Corrosion Inhibitor for Oil & Gas Industries." *International Journal of Biological Macromolecules* 144 (2020): 305-15.

Ansari, Kashif Rahmani, Sudheer, Ambrish Singh, and Mumtaz Ahmad Quraishi. "Some Pyrimidine Derivatives as Corrosion Inhibitor for Mild Steel in Hydrochloric Acid." *Journal of Dispersion Science and Technology* 36, no. 7 (2015): 908-17.

Anusuya, Nagarajan, Jagadeesan Saranya, Palanisamy Sounthari, Abdelkader Zarrouk, and Subramanian Chitra. "Corrosion Inhibition and Adsorption Behaviour of Some Bis-Pyrimidine Derivatives on Mild Steel in Acidic Medium." *Journal of Molecular Liquids* 225 (2017): 406-17.

Baboian, Robert. *Corrosion Tests and Standards: Application and Interpretation*. ASTM international, 2005. Copyright © ASTM, 100 Barr Harbor Drive, West Conshohocken, PA 19428-2959, United States.

Bazgir, Ayoob, Maryam Mohammadi Khanaposhtani, and Ali Abolhasani Soorki. "One-Pot Synthesis and Antibacterial Activities of Pyrazolo [4', 3': 5, 6] Pyrido [2, 3-D] Pyrimidine-Dione Derivatives." *Bioorganic & Medicinal Chemistry Letters* 18, no. 21 (2008): 5800-03.

Chaubey, Namrata, Ahsanulhaq Qurashi, Dheeraj Singh Chauhan, and MA Quraishi. "Frontiers and Advances in Green and Sustainable Inhibitors for Corrosion Applications: A Critical Review." *Journal of Molecular Liquids* (2020): 114385. https://doi.org/10.1016/j.molliq.2020.114385.

Chauhan, Dheeraj Singh, Ahmad Asad Sorour, and Mumtaz Ahmad Quraishi. "An Overview of Expired Drugs as Novel Corrosion Inhibitors for Metals and Alloys." *International Journal of Chemistry and Pharmaceutical Sciences* 4, no. 12 (2016): 680–91.

Chauhan, Dheeraj Singh, Khadija EL Mouaden, Mumtaz Ahmad Quraishi, and Lahcen Bazzi. "Aminotriazolethiol-Functionalized Chitosan as a Macromolecule-Based Bioinspired Corrosion Inhibitor for Surface Protection of Stainless Steel in 3.5% Nacl." *International Journal of Biological Macromolecules* 152 (2020): 234–41.

Chauhan, Swati, Pratibha Verma, Ankush Mishra, and Vandana Srivastava. "An Expeditious Ultrasound-Initiated Green Synthesis of 1, 2, 4-Thiadiazoles in Water." *Chemistry of Heterocyclic Compounds* 56, no. 1 (2020): 123-26.

Cottis, RA. *Shreir's Corrosion.* 2010.

Dhillon, Shweta, and Rama Kant. "Theory for Electrochemical Impedance Spectroscopy of Heterogeneous Electrode with Distributed Capacitance and Charge Transfer Resistance." *Journal of Chemical Sciences* 129, no. 8 (2017): 1277-92.

Dohare, Parul, Dheeraj Singh Chauhan, Belkheir Hammouti, and Mumtaz Ahmad Quraishi. "Experimental and Dft Investigation on the Corrosion Inhibition Behavior of Expired Drug Lumerax on Mild Steel in Hydrochloric Acid ". *Analytical and Bioanalytical Electrochemistry* 9 (2017): 762.

Dohare, Parul, Dheeraj Singh Chauhan, and Mumtaz Ahmad Quraishi. "Expired Podocip Drug as Potential Corrosion Inhibitor for Carbon Steel in Acid Chloride Solution." *International Journal of Corrosion and Scale Inhibition* 7, no. 1 (2018): 25-37.

Dohare, Parul, Dheeraj Singh Chauhan, Ahmad A Sorour, and Mumtaz Ahmad Quraishi. "Dft and Experimental Studies on the Inhibition Potentials of Expired Tramadol Drug on Mild Steel Corrosion in Hydrochloric Acid." *Materials Discovery* 9 (2017): 30-41.

Eicher, Theophil, Siegfried Hauptmann, and Andreas Speicher. *The Chemistry of Heterocycles: Structures, Reactions, Synthesis, and Applications.* John Wiley & Sons, 2013.

EL Mouaden, Khadija, Dheeraj Singh Chauhan, Mumtaz Ahmad Quraishi, and Lahcen Bazzi. "Thiocarbohydrazide-Crosslinked Chitosan as a Bioinspired Corrosion Inhibitor for Protection of Stainless Steel in 3.5% Nacl." *Sustainable Chemistry and Pharmacy* 15 (2020): 100213.

El Mouaden, Khadija, Brahim El Ibrahimi, Rachid Oukhrib, Lahcen Bazzi, Belkheir Hammouti, Omar Jbara, Ahmed Tara, Dheeraj Singh Chauhan, and Mumtaz Ahmad Quraishi. "Chitosan Polymer as a Green Corrosion Inhibitor for Copper in Sulfide-Containing Synthetic Seawater." *International Journal of Biological Macromolecules* 119 (2018): 1311-23.

ElBelghiti, M, Y Karzazi, A Dafali, B Hammouti, F Bentiss, IB Obot, I Bahadur, and EE Ebenso. "Experimental, Quantum Chemical and Monte Carlo Simulation Studies of 3, 5-Disubstituted-4-Amino-1, 2, 4-Triazoles as Corrosion Inhibitors on Mild Steel in Acidic Medium." *Journal of Molecular Liquids* 218 (2016): 281-93.

Frankel, Gerald S, and Michael Rohwerder. "Electrochemical Techniques for Corrosion." *Encyclopedia of Electrochemistry: Online* (2007).

Haque, Jiyaul, Kashif Rahmani Ansari, Vandana Srivastava, Mumtaz Ahmad Quraishi, and IB Obot. "Pyrimidine Derivatives as Novel Acidizing Corrosion Inhibitors for N80 Steel Useful for Petroleum Industry: A Combined Experimental and Theoretical Approach." *Journal of Industrial and Engineering Chemistry* 49 (2017): 176-88.

Jorcin, Jean-Baptiste, Mark E Orazem, Nadine Pébère, and Bernard Tribollet. "Cpe Analysis by Local Electrochemical Impedance Spectroscopy." *Electrochimica Acta* 51, no. 8-9 (2006): 1473-79.

Joshi, Prathamesh G, Dheeraj S Chauhan, Vandana Srivastava, and Mumtaz Ahmad Quraishi. "Curcumin Decorated Silver Nanoparticles as Bioinspired Corrosion Inhibitor for Carbon Steel." *Current Nanoscience* 18, no. 2 (2022): 266-75. https://doi.org/https://doi.org/10.2174/1573413716666201215170101.

Joule, John A, and Keith Mills. *Heterocyclic Chemistry at a Glance*. John Wiley & Sons, 2012.

Kappe, C Oliver. "Controlled Microwave Heating in Modern Organic Synthesis." *Angewandte Chemie International Edition* 43, no. 46 (2004): 6250-84.

Kappe, C Oliver. "Microwave Dielectric Heating in Synthetic Organic Chemistry." *Chemical Society Reviews* 37, no. 6 (2008): 1127-39.

Katritzky, Alan R, Christopher A Ramsden, John A Joule, and Viktor V Zhdankin. *Handbook of Heterocyclic Chemistry*. Elsevier, 2010.

Kaya, Savaş, Cemal Kaya, Lei Guo, Fatma Kandemirli, Burak Tüzün, İlkay Uğurlu, Loutfy H Madkour, and Murat Saraçoğlu. "Quantum Chemical and Molecular Dynamics Simulation Studies on Inhibition Performances of Some Thiazole and Thiadiazole Derivatives against Corrosion of Iron." *Journal of Molecular Liquids* 219 (2016): 497-504.

Khaled, KF. "Application of Electrochemical Frequency Modulation for Monitoring Corrosion and Corrosion Inhibition of Iron by Some Indole Derivatives in Molar Hydrochloric Acid." *Materials Chemistry and Physics* 112, no. 1 (2008): 290-300.

Kimura, Takahide. "Application of Ultrasound to Organic Synthesis." Chap. 171 In *Sonochemistry and the Acoustic Bubble*, 171-86: Elsevier, 2015.

Korde, Rishi, Chandra Bhan Verma, EE Ebenso, and MA Quraishi. "Electrochemical and Thermo Dynamical Investigation of 5-Ethyl 4-(4-Methoxyphenyl)-6-Methyl-2-Thioxo-1, 2, 3, 4 Tetrahydropyrimidine-5-Carboxylate on Corrosion Inhibition Behavior of Aluminium in 1m Hydrochloric Acid Medium." *International Journal of Electrochemical Science* 10 (2015): 1081-93.

Kumar, BR Prashantha, Gopu Sankar, RB Nasir Baig, and Srinivasan Chandrashekaran. "Novel Biginelli Dihydropyrimidines with Potential Anticancer Activity: A Parallel Synthesis and Comsia Study." *European journal of medicinal chemistry* 44, no. 10 (2009): 4192-98.

Lagoja, Irene M. "Pyrimidine as Constituent of Natural Biologically Active Compounds." *Chemistry & Biodiversity* 2, no. 1 (2005): 1-50.

McCafferty, E. "Validation of Corrosion Rates Measured by the Tafel Extrapolation Method." *Corrosion Science* 47, no. 12 (2005): 3202-15.

Mouaden, Khadija EL, Dheeraj Singh Chauhan, MA Quraishi, Lahcen Bazzi, and Mustapha Hilali. "Cinnamaldehyde-Modified Chitosan as a Bio-Derived Corrosion Inhibitor for Acid Pickling of Copper: Microwave Synthesis, Experimental and Computational Study." *International Journal of Biological Macromolecules* 164 (2020): 3709-17.

Obot, IB, K Haruna, and TA Saleh. "Atomistic Simulation: A Unique and Powerful Computational Tool for Corrosion Inhibition Research." *Arabian Journal for Science and Engineering* 44, no. 1 (2019): 1-32.

Obot, IB, DD Macdonald, and ZM Gasem. "Density Functional Theory (Dft) as a Powerful Tool for Designing New Organic Corrosion Inhibitors. Part 1: An Overview." *Corrosion Science* 99 (2015): 1-30.

Obot, IB, and Ikenna B Onyeachu. "Electrochemical Frequency Modulation (Efm) Technique: Theory and Recent Practical Applications in Corrosion Research." *Journal of Molecular Liquids* 249 (2018): 83-96.

Onyeachu, Ikenna B, Ime Bassey Obot, Aeshah H Alamri, and Chinenye A Eziukwu. "Effective Acid Corrosion Inhibitors for X60 Steel under Turbulent Flow Condition Based on Benzimidazoles: Electrochemical, Theoretical, Sem, Atr-Ir and Xps Investigations." *The European Physical Journal Plus* 135, no. 1 (2020): 129.

Polshettiwar, Vivek, and Rajender S Varma. "Microwave-Assisted Organic Synthesis and Transformations Using Benign Reaction Media." *Accounts of Chemical Research* 41, no. 5 (2008): 629-39.

Priyanka, D, S Vinuchakravarthi, D Nalini, MA Quraishi, and DS Chauhan. "Graphene Oxide Integrated into Protective Coatings against Corrosion for Metals and Its Alloys: A Review." *International Journal of Corrosion and Scale Inhibition* 11, no. 2 (2022): 478-506.

Quraishi, MA, and Dheeraj Singh Chauhan. "Environmentally Sustainable Corrosion Inhibitors in Oil and Gas Industry." Chap. 10 In *Organic Corrosion Inhibitors: Synthesis, Characterization, Mechanism, and Applications*, edited by Chandra Bhan Verma, Chaudhery Mustansar Hussain and Eno E. Ebenso, 221-40: John Wiley & Sons, Inc., 2022.

Quraishi, MA, and Dheeraj Singh Chauhan. "Recent Trends in the Development of Corrosion Inhibitors." In *A Treatise on Corrosion Science, Engineering and Technology*, edited by U. Kamachi Mudali, Toleti Subba Rao, S. Ningshen, Pillai Radhakrishna G., Rani P. George and T. M. Sridhar, 783-99: Springer, 2022.

Quraishi, MA, Dheeraj Singh Chauhan, and Aisha Farhat Ansari. "Development of Environmentally Benign Corrosion Inhibitors for Organic Acid Environments for Oil-Gas Industry." *Journal of Molecular Liquids* 329 (2021): 115514. https://doi.org/10.1016/j.molliq.2021.115514.

Quraishi, Mumtaz A, Dheeraj S Chauhan, and Viswanathan S Saji. "Heterocyclic Biomolecules as Green Corrosion Inhibitors." *Journal of Molecular Liquids* (2021): 117265.

Quraishi, Mumtaz Ahmad, Dheeraj Singh Chauhan, and Viswanathan S. Saji. *Heterocyclic Organic Corrosion Inhibitors: Principles and Applications*. Elsevier Inc. Amsterdam, 2020. Book. doi:10.1016/B978-0-12-818558-2.00001-1.

Revie, R Winston. *Uhlig's Corrosion Handbook*. Vol. 57: John Wiley & Sons, 2011.

Reznik, VS, VD Akamsin, Yu P Khodyrev, RM Galiakberov, Yu Ya Efremov, and L Tiwari. "Mercaptopyrimidines as Inhibitors of Carbon Dioxide Corrosion of Iron." *Corrosion Science* 50, no. 2 (2008): 392-403.

Sastri, Vedula S. *Corrosion Inhibitors: Principles and Applications.* John Wiley & Sons, 1998.

Singh, Ashish K, Sudhish K Shukla, and Mumtaz Ahmad Quraishi. "Ultrasound Mediated Green Synthesis of Hexa-Hydro Triazines." *Journal of Materials and Environmental Science* 2, no. 4 (2011): 403-06.

Singh, Priyanka, Dheeraj Singh Chauhan, Sampat Singh Chauhan, Gurmeet Singh, and Mumtaz Ahmad Quraishi. "Bioinspired Synergistic Formulation from Dihydropyrimdinones and Iodide Ions for Corrosion Inhibition of Carbon Steel in Sulphuric Acid." *Journal of Molecular Liquids* 298 (2019): 112051.

Singh, Priyanka, Dheeraj Singh Chauhan, Sampat Singh Chauhan, Gurmeet Singh, and Mumtaz Ahmad Quraishi. "Chemically Modified Expired Dapsone Drug as Environmentally Benign Corrosion Inhibitor for Mild Steel in Sulphuric Acid Useful for Industrial Pickling Process." *Journal of Molecular Liquids* 286 (2019): 110903.

Singh, Priyanka, Dheeraj Singh Chauhan, Kritika Srivastava, Vandana Srivastava, and Mumtaz Ahmad Quraishi. "Expired Atorvastatin Drug as Corrosion Inhibitor for Mild Steel in Hydrochloric Acid Solution." *International Journal of Industrial Chemistry* 8, no. 4 (2017): 363-72.

Singh, Priyanka, Ambrish Singh, and Mumtaz Ahmad Quraishi. "Thiopyrimidine Derivatives as New and Effective Corrosion Inhibitors for Mild Steel in Hydrochloric Acid: Electrochemical and Quantum Chemical Studies." *Journal of the Taiwan Institute of Chemical Engineers* 60 (2016): 588-601.

Tierney, Jason, and Pelle Lidström. *Microwave Assisted Organic Synthesis.* John Wiley & Sons, 2009.

Trivedi, Amit R, Vimal R Bhuva, Bipin H Dholariya, Dipti K Dodiya, Vipul B Kataria, and Viresh H Shah. "Novel Dihydropyrimidines as a Potential New Class of Antitubercular Agents." *Bioorganic & medicinal chemistry letters* 20, no. 20 (2010): 6100-02.

Verma, Chandrabhan, Jiyaul Haque, Mumtaz Ahmad Quraishi, and Eno E Ebenso. "Aqueous Phase Environmental Friendly Organic Corrosion Inhibitors Derived from One Step Multicomponent Reactions: A Review." *Journal of Molecular Liquids* 275 (2018): 18-40.

Verma, Chandrabhan, LO Olasunkanmi, IB Obot, Eno E Ebenso, and Mumtaz Ahmad Quraishi. "5-Arylpyrimido-[4, 5-B] Quinoline-Diones as New and Sustainable Corrosion Inhibitors for Mild Steel in 1 M Hcl: A Combined Experimental and Theoretical Approach." *RSC Advances* 6, no. 19 (2016): 15639-54.

Verma, Chandrabhan, Lukman O Olasunkanmi, Eno E Ebenso, Mumtaz Ahmad Quraishi, and IB Obot. "Adsorption Behavior of Glucosamine-Based, Pyrimidine-Fused Heterocycles as Green Corrosion Inhibitors for Mild Steel: Experimental and Theoretical Studies." *The Journal of Physical Chemistry C* 120, no. 21 (2016): 11598-611.

Verma, Chandrabhan, MA Quraishi, K Kluza, M Makowska-Janusik, Lukman O Olasunkanmi, and Eno E Ebenso. "Corrosion Inhibition of Mild Steel in 1m Hcl by

D-Glucose Derivatives of Dihydropyrido [2, 3-D: 6, 5-D′] Dipyrimidine-2, 4, 6, 8 (1h, 3h, 5h, 7h)-Tetraone." *Scientific Reports* 7 (2017): 44432.

Xiong, Sang, Dong Liang, Zhixin Ba, Zhen Zhang, and Shuai Luo. "Adsorption Behavior of Thiadiazole Derivatives as Anticorrosion Additives on Copper Oxide Surface: Computational and Experimental Studies." *Applied Surface Science* 492 (2019): 399-406.

Xu, Bin, Wenzhong Yang, Ying Liu, Xiaoshuang Yin, Weinan Gong, and Yizhong Chen. "Experimental and Theoretical Evaluation of Two Pyridinecarboxaldehyde Thiosemicarbazone Compounds as Corrosion Inhibitors for Mild Steel in Hydrochloric Acid Solution." *Corrosion Science* 78 (2014): 260-68.

Yadav, Dileep Kumar, B Maiti, and Mumtaz Ahmad Quraishi. "Electrochemical and Quantum Chemical Studies of 3, 4-Dihydropyrimidin-2 (1h)-Ones as Corrosion Inhibitors for Mild Steel in Hydrochloric Acid Solution." *Corrosion Science* 52, no. 11 (2010): 3586-98.

Yadav, Dileep Kumar, and Mumtaz Ahmad Quraishi. "Application of Some Condensed Uracils as Corrosion Inhibitors for Mild Steel: Gravimetric, Electrochemical, Surface Morphological, Uv–Visible, and Theoretical Investigations." *Industrial & Engineering Chemistry Research* 51, no. 46 (2012): 14966-79.

Index

A

acid, viii, 22, 61, 63, 64, 75, 79, 81, 86, 88, 89, 94, 95, 96, 97, 98, 99, 101, 106, 107, 109, 110, 111, 112, 113, 114, 117, 118, 119, 122, 124, 126, 127, 128, 135, 146, 149, 150, 151, 152, 153, 154
adsorption, ix, 131, 132, 133, 135, 139, 140, 142, 143, 144, 145, 147, 148, 149, 153, 154
alkaloids, 58, 59, 60, 61, 62, 63, 64, 65, 66, 67, 68, 69, 70, 72, 73, 74, 87, 93, 95, 96, 97, 98, 99, 100, 101
antibiotic(s), 67, 81, 83, 86, 88, 95, 97, 98, 99, 100, 101, 127
antifungal activity, v, vii, 1, 2, 3, 4, 17, 51, 52, 69, 111, 127, 128
Apelblat model, 11, 26

B

bacterium, v, vii, 109, 110, 111, 126, 128
biosynthesis, viii, 53, 83, 88, 95, 97, 109, 110, 121, 123, 126, 127, 128, 129, 130

C

cell extract preparation, 114
corrosion, v, vii, ix, 131, 132, 133, 134, 135, 136, 138, 139, 142, 143, 144, 145, 146, 148, 149, 150, 151, 152, 153, 154
corrosion inhibitor, v, vii, ix, 131, 132, 133, 134, 135, 136, 138, 140, 143, 144, 145, 146, 148, 149, 150, 151, 152, 153, 154

D

diazines, 132, 133, 134

differential scanning calorimeter (DSC), vii, 2, 9, 11, 18, 19, 21, 24, 51
dihydro-oxipinopyrazinopyrimidine alkaloids, 73
dihydro-pyrimidine alkaloids, 66

E

enzymes assays, 114
equilibrium solubility, 10, 19, 21, 34, 37

F

fused pyrimidine alkaloids, 69

G

Gibbs energy, 13, 14, 15, 27, 28, 32, 38, 40, 50, 52

H

Hansen solubility parameters (HSP), 40, 43, 44, 52
heterocycles, 3, 4, 23, 58, 107, 132, 134, 135, 137, 140, 142, 146, 148, 150, 153
hybrids, v, vii, 1, 2, 3, 15, 16, 17, 53, 105, 106, 107

K

kinetic solubility, 10, 19, 20, 24, 34, 35, 51

L

lipophilicity, vii, 2, 3, 4, 29, 31, 34, 48, 54, 59

M

modified Apelblat equation, vii, 2, 12, 48, 52

N

natural products, v, vii, viii, 57, 58, 59, 60, 72, 93, 94, 95, 98, 99, 100, 101, 102, 128
nucleosides/nucleotides, 74, 77, 78, 80

O

orotic, viii, 109, 110, 113, 115, 117, 118, 124

P

powder X-ray diffraction (PXRD), 10
protein assay, 116
pseudomonad, 110, 111, 112, 124, 125
Pseudomonas chlororaphis (*P. chlororaphis*), v, vii, viii, 109, 111, 112, 117, 120, 121, 123, 127, 128
purine, 58, 78, 80, 81, 122
PXRD method, 18
pyrimidine, v, vii, viii, ix, 3, 4, 5, 17, 18, 19, 21, 23, 29, 57, 58, 59, 60, 63, 64, 65, 66, 68, 72, 73, 74, 76, 77, 78, 81, 82, 84, 85, 86, 88, 89, 91, 92, 93, 95, 96, 97, 98, 102, 105, 109, 110, 112, 113, 114, 116, 117, 118, 119, 120, 121, 123, 124, 126, 127, 128, 129, 130, 131, 132, 134, 135, 136, 137, 138, 140, 142, 144, 145, 146, 148, 149, 151, 153
pyrimidine derivatives, ix, 3, 4, 18, 19, 21, 23, 29, 105, 131, 133, 135, 136, 138, 140, 142, 144, 145, 146, 148, 149, 151
pyrimidine ring, v, vii, viii, 57, 59, 60, 66, 73, 76, 93, 135, 144, 148

pyrimidines, v, vii, viii, ix, 1, 15, 16, 17, 51, 53, 54, 57, 97, 98, 110, 112, 114, 119, 125, 130, 131, 132, 133, 134, 135, 148

Q

quinazoline alkaloids, 69, 71, 99

S

solubility, vii, 2, 3, 4, 10, 11, 12, 13, 19, 20, 21, 22, 23, 24, 25, 26, 27, 31, 33, 34, 35, 36, 37, 38, 40, 41, 43, 44, 45, 46, 47, 48, 50, 51, 52, 53, 55, 140
substituted-pyrimidine alkaloids, 60
synthesis, v, vii, ix, 1, 2, 3, 5, 6, 7, 8, 15, 16, 17, 52, 53, 54, 60, 75, 76, 78, 79, 81, 84, 86, 88, 90, 94, 95, 96, 97, 98, 99, 100, 101, 102, 103, 105, 106, 107, 108, 110, 111, 112, 119, 121, 123, 124, 127, 128, 129, 130, 135, 136, 137, 138, 148, 149, 150, 151, 152, 153

T

tethered pyrimidine alkaloids, 60
tetrahydro-pyrimidine alkaloids, 68
thermodynamic functions, viii, 2, 27, 31, 38, 40, 50, 52
thermodynamic solubility, 24, 29, 36, 39
toxin(s), 92

U

uracil, viii, 76, 88, 109, 110, 111, 113, 117, 118, 119, 121, 124, 125, 127, 135

V

vitamin(s), 89, 90, 96, 101, 105